INTRODUCTION TO ERROR ANALYSIS

THE SCIENCE OF MEASUREMENTS, UNCERTAINTIES, AND DATA ANALYSIS

By

Jack Merrin

Institute of Science and Technology Austria
IST Austria
Am Campus 1
3400 Klosterneuburg, Austria
jack.merrin@ist.ac.at

CreateSpace Independent Publishing Platform
Version 1.1

ii

Preface to the first edition

There is a need for more discussion of error analysis in the physical sciences curriculum. This book is appropriate for a laboratory course or actual research. Some people may find it hard to get on the correct path amongst all the probability and statistics books available. Some content is even contradictory. There are only a few books that classify themselves as error analysis. Start with error analysis first, then add in probability, and then some statistics if you wish.

Don't let your reputation be ruined by poor data analysis. The accepted conventions for data analysis rely on calculus. There is no watered down version for algebra students. Maintaining high standards is necessary if science is to progress.

Older error analysis books refer to out of date code in Excel or FORTRAN. I suggest students learn MATLAB at the first opportunity. MATLAB provides confidence and independence to solve or simulate more complex problems. My book will not teach you MATLAB, but I do provide all the example code which you can adapt. I recommend Stormy Attaway's book (4th edition). If one cannot afford MATLAB, try the free program Octave. I use Octave on my new computer and don't notice much difference.

I also recommend Kaleidagraph and Graphpad Prism, but I do not talk about these programs in my book. Kaleidagraph has great and easy to use nonlinear curve fitting capabilities. Graphpad Prism makes the nicest figures. Between all these you should have everything covered. If you don't want to write your own curve fitting programs you can use these programs and know what is under the hood after finishing this book.

It shouldn't cost a fortune to learn error analysis. In the references section, I suggest books you can buy and not waste your money. My idea was to self publish this book in paperback and on Kindle to keep the price low compared to all the other books. If you have comments about my book or suggestions please contact me and jackmerrin@gmail.com. I plan to update the book periodically. Leaving a verified purchase review on Amazon is also helpful to me.

Thanks,
Jack Merrin
August 15, 2017

Contents

Chapter 1

Error analysis introductory concepts

1.1 Making sense of randomness

Randomness is a normal aspect of the physical world. In biology, genetically identical cells can behave differently because because they contain different numbers of molecules. Chemical reactions also occur randomly in space and time. Randomness is built into the theory of quantum mechanics. Do you think the stock market is predictable? Randomness is fundamental to scientific processes, but it doesn't equate to poor understanding.

Often in science, we hear people talking about uncertainty, error, and randomness. This doesn't mean that scientists don't know what they are talking about or make mistakes. Often how we talk about science does not relate to what is actually demonstrated. Have you ever heard, "Well it's just a theory?" or "You can't really prove it for sure." These types of statements don't win debates.

When statements are backed up by repeated measurements and mathematics they hang around longer. You may not know perfectly, but with mathematics you can calculate a probability that something is correct. Many statements in particle physics have the standard of 1 part in a million accuracy. It would be like winning the lottery for one of these results to be false. People who play the lottery also fly on planes. They think they can win 1 in a million, but not lose 1 in a million in a plane crash. How do you view luck and certainty?

More advanced theories expand our knowledge or interpretation, but cannot violate well tested experimental results. The theory of relativity overthrows Newtonian mechanics, but Newtonian results are quite accurate when speeds are small compared to the speed of light. Both theories predict the same numbers to very high precision.

1.2 How error analysis works

The best way to establish a theory is to test it with repeated experiments. Without experimental verification, we are left at best with mathematics or at worse philosophy. One might argue that some things are too simple to measure or have already been measured. Then there would be no purpose in repeating them. In response to that, one might argue science is an art. There are many paintings of still life. That doesn't stop others from painting their own version. Beware, knowledge is the death of research!

The primary function of error analysis is characterization and reduction of the error in measurements and reporting their values and uncertainties. Error analysis is a system in its own right, but cherry picks parts of statistics and probability theory to focus on. In error analysis, one takes care in how values are reported with significant figures and rounding. Error analysis provides a method of calculating errors in derived quantities from known measurements called propagation of uncertainty or propagation of error. Curve fitting in error analysis is used to extract model parameters and their uncertainties. Measurement of quantities from different groups can be combined in error analysis to find a new average and uncertainty via the weighted mean. Error analysis therefore takes care of the fundamental constants and their accepted values.

1.3   How probability works

Probability theory is pure mathematics. Generally, one starts with axioms and derives the entire theory. This is not the case in physics, one has to do experiments to prove things. Probability theory therefore does not usually deal with actual data. It predicts what occurs in a certain situation with certain given starting parameters. In this way, one does not have an inverse problem of deriving a model. The model is set and the consequences are derived.

A certain few probabilistic models are useful to know because they occur frequently in experimental science. For example, exponential functions, Gaussian functions, binomial distributions, and Poisson distributions to name some. These all describe physical phenomenon like radioactive decay, SAT scores, IQ, biased coin flips, and how often you get a certain number of chocolate chips in your cookie. Usually we take measurements first and then try to plug the results into one of the ready made probability models if we think that makes sense.

1.4   How statistics works

Statistics is also mathematics, but more of an applied nature. Statistics hopes to understand why a certain dataset occurs the way it does as an inverse problem. In this sense, it is more similar to error analysis than probability theory. Some definitions in statistics are arbitrary, but the theory seems to hold together as a whole. Statisticians often work with hypothetical data or collaborate with scientists and analyze real datasets. Calculations of all sort of things from data sets occur in statistics, but error analysis is usually only concerned with a few statistics.

1.5   Accuracy and precision

Measurements are said to be accurate if they are centered around the true value. We often don't know what the true value is, like for the mass of an electron, but we do have the accepted value. Measurements are said to be inaccurate if their average is away from the true value. Generally, systematic errors will cause a measurement to be inaccurate. Measurements are said to be precise if the measurements fall within a narrow range. If the measurements are spread over a wide range they are said to be imprecise. In experiments, one strives to make precise and accurate measurements.

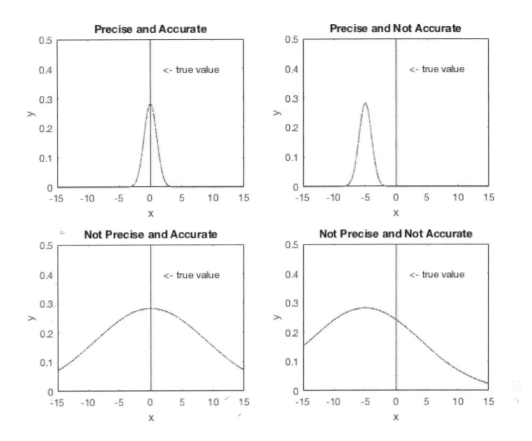

Figure 1.1: Precison and Accuracy

Precision also refers to a measuring device. If you have a volt meter that can measure 0.01V then the precision is 0.010V. It is likely when you repeat a measurement on a such a volt meter you will get the exact same result. You have not hit the fundamental random error yet and need a more sensitive meter to see it. Precision and sensitivity mean the number of digits you can record in a single measurement. Most measuring devices are digital because the act of measurement is to write down a terminating decimal number. If you read a galvanometer with your eye then you are making a digital measurement even though a galvanometer is an analog current measurement device.

Example 1.1. Write a MATLAB code illustrating precision and accuracy

Solution 1.1. This code demonstrates precision as a narrow distribution and accuracy as the average close to the true value.

```
% Code demonstrating precision and accuracy.
clear;
figure(1);clf;
x = -15:0.01:15;
sp = 1;
```

```
snp = 9;
a = 0;
na = -5;
a1 = [ 0 0 ];
a2 = [ 0 0.5];

y1 = 1/(sp*2*sqrt(pi))*exp(-(x-a).^2/(2*sp^2));
y2 = 1/(sp*2*sqrt(pi))*exp(-(x-na).^2/(2*sp^2));
y3 = 1/(sp*2*sqrt(pi))*exp(-(x-a).^2/(2*snp^2));
y4 = 1/(sp*2*sqrt(pi))*exp(-(x-na).^2/(2*snp^2));
subplot(2,2,1)
plot(x,y1);hold on;
plot(a1,a2,'r');
xlabel('x');ylabel('y');title('Precise and Accurate');
axis([-15 15 0 .5]);text( 2, 0.4, '<- true value');

subplot(2,2,2)
plot(x,y2);hold on;
xlabel('x');ylabel('y');title('Precise and Not Accurate');
axis([-15 15 0 .5]);
plot(a1,a2,'r');text( 2, 0.4, '<- true value');

subplot(2,2,3)
plot(x,y3);hold on;\xlabel('x');ylabel('y');
title('Not Precise and Accurate'); axis([-15 15 0 .5]);
plot(a1,a2,'r'); text( 2, 0.4, '<- true value');

subplot(2,2,4)
plot(x,y4);hold on;
xlabel('x');ylabel('y');title('Not Precise and Not Accurate');
axis([-15 15 0 .5]); text( 2, 0.4, '<- true value');
plot(a1,a2,'r');
```

Normal distribution

$$y = \frac{1}{\sigma\sqrt{2\pi}} e^{\frac{(x-\mu)^2}{2\sigma^2}}$$

let $\mu = 0$ ∧ $\sigma = 1$

$$y = \frac{1}{\sqrt{2\pi}} e^{\frac{x^2}{2}}$$

Standard normal distributions

1.6   The nature of random and systematic error

Random error is the normal fluctuation associated with a measurement. By fluctuations, one measures a distribution of values after repeated measurement. Random measurement can be eliminated to some degree by repeating the measurement many times and calculating statistics like the mean and standard deviation of the mean which become more precise with the number of measurements. Random error is described by its precision, how close the measured values cluster together. One tries to understand where the random error comes from and reduce it, but it cannot be eliminated.

   If one is counting photons emitted from a light source, then the measurement can fluctuate when you have small numbers of photons. This fluctuation is proportional to the $\sqrt{N}$. A one Watt laser the fluctuation will be pretty small, but if you measure 100 photons per second you would expect to see different measurements in a range of 90-110 per second. This type of noise is of a fundamental nature. It is how the statistics of photon emission work so you can't reduce it. You can however quantify it, measure the mean accurately, and distinguish 90-110 per second from 95-115 per second.

Random error can also be caused by the apparatus. For example, if you use a laser, then the power of the laser might fluctuate by 1 percent. To reduce this error, you could buy a better laser. You might also try to split the beam and simultaneously measure the power of the laser and the fluorescent light output of the sample. The ratio of the two should be constant and you could therefore eliminate the defect with your apparatus. Some physicists go down deep in the cave in tunnels and build extremely large experiments hoping to measure rare events that are lost in the noise on the surface. They avoid cosmic radiation and vibrations. One can also improve an apparatus sometimes by placing it in a temperature controlled box. This eliminates subtle effects like thermal expansion.

Systematic errors generally shift measurements away from the true value. An example of a systematic error is an uncalibrated thermometer. You read the thermometer properly but the reading is say 0.5 K away from where it should be. No amount of repeated measurement can overcome a systematic error. Millikan measured that charge was quantized, but his value differed from later measurements. Only later, it was determined that he was using the wrong value for viscosity of oil. Once that was corrected, his value agreed with the other measurements.

Systematic error also includes gross mistakes. The Mars Climate Orbiter crashed into Mars. After careful analysis, they found the hardware and software groups were using different units. One was using meters and the other inches. It may be challenging to eliminate systematic errors, but it is necessary to get the true result.

One can correct for systematic error by calibration of the measurement devices. One can try a sample with a known property that can be measured to test the apparatus. For example, if you are measuring the emission spectrum of a certain fluorescent molecule, you could first perform a test with a known substance. This will let you know if your equipment is functioning properly. You can also try the measurement on someone else's machine and see if you get the same result also. By calibrating at some reference condition, you can zero the measuring device and eliminate systematic errors.

All errors are not entirely experimental. Theoretical error is possible. This can happen when a calculation is difficult and one cannot precisely determine a constant in a theory for example. Some theories might be based on simulations and there is random error associated with a simulation just like there is in an experiment.

1.7 Reproducibility

Reproducibility is a very important aspect of science. I don't believe anything until I have measured it at least three times. Does it work tomorrow? Does it work next week? If a measurement is important, it will surely be repeated eventually with attempts to improve upon it. One should describe how one does an experiment with enough detail such that another independent person or lab can attempt to reproduce it. For some in the rush to publish, things can be sloppy or incorrect. In a recent survey, 70 percent of scientists were unable to reproduce someone elses experiment at some point in their career. Half of the people surveyed say this a crisis.

The best way to make your data reproducible is to be rigorous and use error analysis. Don't cherry pick your data. If you use something from someone else, verify it for yourself first before using it. Otherwise you risk not being reproducible. I would say most irreproducible results are not intentional misrepresentation, but just being sloppy. I recall one experiment where they claimed that neutrinos go faster than the speed of light. It turned out the reason was a loose cable. Before you try to overthrow Einstein make sure your cables are plugged in!

1.8　Falsifiabiilty and certainty

Rutherford said that if you have to use statistics to prove your argument you should have done a better experiment. If you come to a conclusion and there is no apparent experimental test of your theory, then your theory is not falsifiable. There is no certainty without the test of falsifiability. A lot of effort goes into string theory, but none of the theories produce experimental tests. Is that the right way to do science?

Most results that are numerical are by nature falsifiable. If I say there is a ten percent chance this drug will cure cancer. Eventually when you try it on enough people you can find out if it really works statistically. Sample size is important in establishing or falsifying a claim. Make sure your claims are falsifiable and do the experiments that would falsify it. Maybe you can discover something new. This usually happens when something is found to be actually false.

1.9　Keeping records

It is easy to lose track of what you are doing unless you keep a proper notebook. Always record the date of your experiments, and keep the pages in sequential order. Never cross out anything in your notebook. If there is a mistake, make an extra note. It used to be that you had to have your notebook signed to established intellectual property. Things have since changed. Depending on where you live, it is no longer first to publish or record in your notebook, but first to file the patent. So keep good notes anyways, but be aware of intellectual property laws.

Write descriptions of experimental settings so they can be reproduced. If possible record a written copy of your data or the name of the data file where the numbers are saved. After you are finished for the day, write a summary and analyze the data and put the graphs. Put the date, title, axes labels, and units on your graph. When you are keeping records in data files you should be careful about your organization. You want to be able to come back five years to the data file and understand what the numbers mean. This might mean you need to insert a text file which describes what the columns of data are.

You also need to include the date in the text file to record when the measurements were taken. Some people use folders with the date of the measurements so you have a record of what you did on what day. Sometimes when you go into the data file and type something then save it then it overwrites the actual date of the file with the current date. Copying the file may adjust the date also. If you don't have the date recorded then you will quickly lose track of what is going on.

Data files can be lost or corrupted. You should back up the data files. If measurements don't involve a lot of numbers, you should just record the numbers manually into your notebook.

1.10 Recording measurements

The precision of your meter limits the precision of your final result. Digital devices like a volt meter give you a certain number of digits you can record. If a measurement of the resistance keeps reading 1.023Ω, you can't gain anything by repeated measurement. If you need more precise results, you have to buy a more expensive meter with more digits of precision. Then you will observe fluctuations in the measurement due to random error. The best digital devices round the last digit. The worst digital devices just chop off the last digit.

Analog devices produce a continuous output. A galvanometer measures electric current. In the old style meters, a little pointer would move on the gauge and you would have to eyeball the result. In effect, you convert an analog display into a digital measurement because you have a finite number of digits in your description. In reality, all measurements are digital.

In the next chapter, you will gain some concept of significant digits. Once you understand this, it will become clear how to record data so you can use it later. It doesn't hurt to record all the digits from your meter. In the end analysis, you may not need them all. The more measurements you take, then the more significant digits you need to keep when calculating statistics.

1.11 NIST, CODATA, and the fundamental constants

CODATA stands for Committee on Data for Science and Technology. They are charged with collecting all the experimental results and computing the accepted values for the fundamental constants and their uncertainties. NIST stands for the National Institute of Standards and Technology. The current fundamental constants and their uncertainties can be found by searching the NIST website. These are the most precise measurements known to man. Usually from year to year, the uncertainty in the values goes down, but it can also go up.

When one looks at the fundamental constants on the NIST website, one is struck by how they are presented. The values are presented in scientific notation with SI units. They designate the uncertainty in the last two digits by enclosing the measurement errors in parenthesis. For all the measurements, the uncertainty is always two significant digits. These measurements are the most professional available so we should try to emulate what they are doing.

An interesting idea is how the measurements of different labs are combined to give the final result and the uncertainty. CODATA uses a convention called the weighted mean which is a least squares technique. You will learn about this in the coming chapters of this book. If you didn't get a chance to see the website, here is a list of some common fundamental constants.

Avogadro's constant	$N_A = 6.022140857(74) \times 10^{23}$ mol^{-1}
Boltzmann's constant	$k_B = 1.38064852(79) \times 10^{-23}$ J K^{-1}
Permittivity of free space	$\epsilon_0 = 8.854187817... \times 10^{-12}$ F m^{-1}
Mass of the electron	$m_e = 9.10938356(11) \times 10^{-31}$ kg
Mass of the neutron	$m_n = 1.674927471(21) \times 10^{-27}$ kg
Mass of the proton	$m_p = 1.672621898(21) \times 10^{-27}$ kg
Electron volt	$eV = 1.6021766208(98) \times 10^{-19}$ J
Charge of the electron	$e = 1.6021766208(98) \times 10^{-19}$ C

Faraday constant	$F = 96485.33289(59) \; \mathrm{C\,mol^{-1}}$
Inverse fine structure constant	$\alpha^{-1} = 137.035999139(31)$
Permeability of free space	$\mu_0 = 4\pi \times 10^{-7} \; \mathrm{N\,A^{-2}}$
Molar gas constant	$R = 8.3144598(48) \; \mathrm{J\,mol^{-1}\,K^{-1}}$
Newton's gravitational constant	$G = 6.67408(31) \times 10^{-11} \; \mathrm{m^3\,kg^{-1}\,s^{-2}}$
Planck's constant	$h = 6.626070040(81) \times 10^{-34} \; \mathrm{J\,s}$
Planck's constant$/(2\pi)$	$\hbar = 1.054571800(13) \times 10^{-34} \; \mathrm{J\,s}$
Rydberg constant	$R_\infty = 10973731.568508(65) \; \mathrm{m^{-1}}$
Speed of light in vacuum	$c = 299792458 \; \mathrm{m\,s^{-1}}$
Stefan Boltzmann constant	$\sigma = 5.670367(13) \times 10^{-8} \; \mathrm{W\,m^{-2}\,K^{-4}}$

One might also notice that some constants do not have an error bar. This is not a typo. Those values are defined so they have zero uncertainty. Examples of constants that are defined are c, ϵ_0, and μ_0. These definitions make it possible to more accurately define the SI units. In the future, other constants like Planck's constant may be defined leading to further accuracy in measurements. One might find a constant like the inverse fine structure constant which has 12 significant digits and wonder if the uncertainty can be reduced further. We will have to wait and see, but some of the most precise atomic clock measurements can reach one part in 10^{17} and these clocks are always being refined to be more precise.

One might wonder why it is necessary to know the fundamental constants so precise. In physics, more encompassing theories are sought out. The goal is that someday the constants can be related by some formula that predicts their values to a high degree. If we know the constants to high accuracy, we can see if the theory agrees or refutes it.

Another idea is that somehow the fundamental constants vary over time or at different places in the universe. If we know the constants now on earth, we can compare astronomy measurements from back in time or see if the values drift on earth. As far as we know, the constants do not change over time and the laws of physics are the same in different parts of the universe.

Notice that the fundamental constants end in two digits of uncertainty for every measurement. This is what the professionals are doing so you should also adopt this rule. It turns out that if you use just one digit you can have significant rounding errors when you calculate uncertainties. It might be as much as fifty percent in the last digit as you go from 0.1 to 0.15. If you use two digits it is only as much as one part in two hundred, the rounding error 0.10 to 0.105. This is the rule of thumb I always use. Also don't use three digits of uncertainty because it looks strange. You can't really justify statistically that many digits so don't write 1.234 ± 0.345. Just write 1.23 ± 0.35.

Another important aspect of error analysis is the units you use to describe your measurements. By default, you should use SI units unless you have a specific reason to use something else. USA and England are two countries that use a mixture of different units. A lot of people are not really informed on what they are. There are exactly 2.54 cm in an inch. That is the definition of imperial length based on SI units. There is no foot-long bar to calibrate inches and feet.

Chapter 2

How to calculate measurements and uncertainties

2.1 The best estimate and the uncertainty

The goal of this chapter is how to calculate the best estimate and its uncertainty. It is important that you calculate an uncertainty. Without an uncertainty a measurement is useless. It cannot be combined with other known facts to produce new facts or hypotheses in a quantitative way.

2.2 Significant digits

2.2.1 Significant digit rules

Many websites and books conflict on the description of significant digits. These are the significant digit rules.

- All non-zero digits are significant.

- All zeroes between non-zero digits are significant.

- Leading zeroes are not significant.

- Trailing zeroes after the decimal point are significant.

- Trailing zeroes without a decimal point are not significant.

Example 2.1. How many significant digits are in the following numbers 2.30, 230, 2300, 23, 3904, π, and c the speed of light in a vacuum.

Solution 2.1. 2.30 has 3 significant digits. 230 has 2 significant digits. 2300 has 2 significant digits. 23 has 2 significant digits. 3904 has 4 significant digits. π has an infinite number of significant digits. c is like π because it is defined exactly as

$$c = 299792458 \, \text{m/s}$$

2.2.2   How to round

A general rule of thumb is to round the uncertainty to two significant digits. First, identify what the rounding digit is. Do you want to round to tenths or thousandths and so forth? Look at the next digit. If it is 5 or more then round up to rounding digit and drop all the rest of the digits to the right. Nothing advanced here.

Here are some examples of rounding correctly to two significant digits.

- $0.0445 \rightarrow 0.045$

- $0.0444 \rightarrow 0.44$

- $12.2 \rightarrow 12$

- $12.9 \rightarrow 13$

- $2501 \rightarrow 2500$

- $2555 \rightarrow 2600$

- $0.999 \rightarrow 1.0$

- $99.64 \rightarrow 100$ In this example, we are really trying to break the significant digit rules. The rules state 100 has one significant digit not two. There are other notations like parenthesis or scientific notation which can better represent the significant digits you want to use.

Here are some examples of incorrectly reported measurements.

- 1.54 ± 1.3

- 1.5 ± 0.43

- These are incorrect because the precision doesn't match. I would be suspicious of the paper or book of anyone who makes this type of error.

- 1.5 ± 0.1 is incorrect because you need two significant digits of uncertainty. Experts always use two significant digits of uncertainty.

- 1.541 ± 0.131 is incorrect.

Here are some examples of correctly reported measurements.

- 1.5 ± 1.3

- 1.50 ± 0.10

2.2.3   Rounding the uncertainty to two digits

Why round uncertainty to two significant digits? Basically, it will help get rid of round off errors in the uncertainty to one part in a hundred. For me, there is only one significant digit rule.

- Round to two significant digits of uncertainty and match the precision of the measurement.

2.2.4 Rounding the measurement to match the uncertainty

I will now demonstrate the fallacy of the original significant digit rules. For example, consider 4200 ± 43 and 4200 ± 1500. In the first case, 4200 must have four significant digits. In the second case, 4200 must have two significant digits. It depends on what the uncertainty is.

2.3 Notation for recording measurements and uncertainty

2.3.1 Plus and minus notation

To apply two significant digits in the uncertainty, you always calculate the uncertainty first. Then you match the precision of the measurement. If your measurement reads $y = 4.221242\ldots$ cm and you calculate the uncertainty of $0.43412\ldots$ cm using some error analysis formula, then we should write

$$y = 4.22 \pm 0.43 \,\text{cm}$$

2.3.2 Parenthesis notation

Parenthesis notation is an alternative to plus or minus notation.

$$y = 4.22(43) \,\text{cm}$$

The numbers in parenthesis represent the uncertainty in the last two digits of y.

2.3.3 Scientific notation

We are not going to write 4200 ± 43 as $420\overline{0} \pm 43$. We also wouldn't write 4200 ± 1500 as $4200(1500)$. Scientific notation is another alternative to represent significant digits.

In scientific notation, we write $A \times 10^B$ where all the digits of A are significant, $1 \le A < 10$, and B is an integer. So we could write

$$4200 \pm 1500 \to 4.2(15) \times 10^3$$

Example 2.2. Write 165.34 and 0.0000412 in scientific notation.

Solution 2.2. For the first number, we just move the decimal point two places to the left.

$$165.34 = 1.6534 \times 10^2$$

For the second number, we move the decimal five places to the right.

$$0.0000412 = 4.12 \times 10^{-5}$$

name	symbol	base 10
yotta	Y	10^{24}
zetta	Z	10^{21}
exa	E	10^{18}
peta	P	10^{15}
tera	T	10^{12}
giga	G	10^{9}
mega	M	10^{6}
kilo	k	10^{3}
milli	m	10^{-3}
micro	μ	10^{-6}
nano	n	10^{-9}
pico	p	10^{-12}
femto	f	10^{-15}
atto	a	10^{-18}
zepto	z	10^{-21}
yocto	y	10^{-24}

2.3.4   Engineering notation

Another alternative to scientific notation is to use engineering notation. If the power of the scientific notation is between 10^{24} and 10^{-24} you can use these common factors like kilo or nano. Scientists often get a feel of the length scale they are working on using engineering notation. Try to keep the numbers in engineering notation between 1 and 999. I wouldn't write 10,000 nm, instead it is better to just write 10 μm.

2.4   The primary statistics

We now show how to calculate the three primary statistics in error analysis: the mean, the standard deviation, and the standard deviation of the mean. There are hundreds of different statistics you can calculate, but these primary statistics will serve you best in the process of error analysis.

2.4.1   The arithmetic mean

The mean is the average of a set of measurements. Imagine N experimental measurements.

$$y_1, y_2, \ldots, y_N$$

These measurements could be the same, different, or described by different numbers of digits. Then the mean is calculated as

$$\overline{y} = \frac{1}{N} \sum_{i=1}^{N} y_i$$

The calculation of $\overline{y}$ can land on an infinite string of digits like $1/3 = 0.3333\ldots$ but we eventually round it.

2.4.2 The standard deviation

The spread in the measurements is characterized by the variance or the standard deviation. The variance is defined as

$$\sigma^2 = \frac{1}{N} \sum_{i=1}^{N} (y_i - \bar{y})^2$$

The standard deviation, σ, is the square root of the variance.

2.4.3 The standard deviation of mean

Since we take a lot of measurements, we know the mean more precisely than just σ. We will show this later in a simulation. A better way to represent the uncertainty is called the standard deviation of the mean.

$$\bar{\sigma} = \frac{\sigma}{\sqrt{N}}$$

When you report the mean and its uncertainty you should generally quote the standard deviation of the mean. The standard deviation doesn't change much if you include more measurements. The reward for taking more measurements is a smaller standard deviation of the mean.

2.4.4 Fractional uncertainty

Fractional uncertainty is a way to compare the quality of different measurements. The lower the fractional uncertainty the better. The fractional uncertainty or relative uncertainty is defined as the uncertainty divided by the mean value. For example, if you have $x = 1.345(22)$ then the fractional uncertainty is $\sigma_x/x = 0.022/1.345 = 0.16$. The best measurements of the fundamental constants have a fractional uncertainty of about one part in a billion. For example, the mass of the electron is

$$m_e = 9.10938356(11) \times 10^{-31} \, \text{kg}$$

The relative uncertainty is

$$\frac{\bar{\sigma}_{m_e}}{\bar{m}_e} = 1.2 \times 10^{-8}$$

Not a bad measurement!

2.5 Calculating the primary statistics

2.5.1 Converting to SI units

It is important to use SI units when you report your results. Suppose you have a meter that reports $1.23456(22)$ inches. The conversion factor is 1 inch is 25.4 mm. So you can write.

$$0.123456(22) \, \text{in} \times \frac{25.4\text{mm}}{\text{in}} = 31.3578(56) \, \text{mm}$$

You might think you can't justify that much precision because the conversion factor only has three significant digits. Imperial units are defined by exact conversion factors to SI. If your conversion factor is some other lousy unit or approximation then you can't justify its exactness.

2.5.2   Test of reporting measurement and uncertainty

Here are some exercises to test your rounding skills and reporting of measurements.

Example   2.3. Round   the   following   numbers   to   2   significant   digits:
$3903, 0.0234, 0.00004503, 100, 1000, 2^{10}, 7!$.

Solution 2.3. $3903 \rightarrow 3.9 \times 10^3$. $0.0234 \rightarrow 0.023$. $0.00004503 \rightarrow 0.000045$, $100 \rightarrow 1.0 \times 10^2$.
$1000 \rightarrow 1.0 \times 10^3$. $2^{10} = 1024 \rightarrow 1.0 \times 10^3$. $7! = 5040 \rightarrow 5.0 \times 10^3$.

Example 2.4. A measurement is given as 4303.125 ± 0.159. Round the uncertainty to two significant digits then round the measurements to match this precision. Repeat for 0.02966 ± 0.00135.

Solution 2.4. For the 1st part, the uncertainty is 0.16. Then you round the measurement to 4303.13. So you report 4303.13(16). For the 2nd part. The uncertainty is 0.0014. The measurement needs to be rounded to 0.0297. So we report 0.0297(14).

Example 2.5. Which of the following expressions is reported correctly? 103 ± 12, 20.3 ± 12, 10.1 ± 0.1, $1.07(12)$, $109(7)$, 314520 ± 2300. Which can be fixed?

Solution 2.5. A correct expression has two significant digits in the uncertainty. 103 ± 12 is ok. 20.3 ± 12 is not ok. 10.1 ± 0.1 is not ok. $1.07(12)$ is ok. $109(7)$ is not ok. 314520 ± 2300 is not ok. Let's fix some of these up. 20.3 ± 12 can be written as $20(12)$. 10.1 ± 0.1 can be written as $10.10(10)$. $109(7)$ can be written as $109.0(70)$. 314520 ± 2300 can be written as $3.145(23) \times 10^5$.

2.5.3   Primary statistics in MATLAB

Example 2.6. A sprinter runs the 100 meter dash and his times are measured as: 10.23 s, 9.99 s, 10.34 s, 10.00 s, and 9.89 s. Find the mean time, the variance, the standard deviation, and the standard deviation of the mean. What do you report?

Solution 2.6. Let's do this in MATLAB

```
t = [10.23 9.99 10.34 10.00 9.89]
m = sum(t)/length(t)
sigma2= sum((t-m).^2)./length(t)
sigma = sqrt(sigma2)
sdom = sigma/sqrt(length(t))
```

The mean time is $\bar{t} = 10.090$ s. The variance is $\sigma^2 = 0.028$ s^2. The standard deviation is $\sigma = 0.17$ s. The standard deviation of the mean is 0.075 s. The average race time is 10.090(75) s.

Example 2.7. Illustrate in a MATLAB program how the mean, standard deviation, and standard deviation of the mean scale with the number of measurements.

Solution 2.7. Here we show that the mean and standard deviation get more precise the more data points you add. The mean and standard deviation approach a constant, but the standard deviation of the mean decreases like $1/\sqrt{N}$.

```
% MATLAB code, dependence of statistics on the number of data points
clear;format long;
N = logspace(1,6,100);

for i = 1:length(N)
    i % Keep track of the progress of the program
    for j = 1:N(i)
        y(j) = normrnd(0,1) + 5; % Gaussian random data mean 5 and ssigma 1.
    end

    m(i) = sum(y)/length(y); % Calculating the mean
    stnd(i) =  sqrt(  1./N(i).* sum( ( y - m(i) ).^2 )); % Calculating sigma
    sdm(i) = stnd(i)/sqrt(N(i)); % Calculating sigma/N^0.5
end

figure(1);clf;
subplot(3,1,1)
loglog(N, m);
xlabel('Number of measurements');ylabel('Mean');
subplot(3,1,2);
loglog(N, stnd);
xlabel('Number of measurements');ylabel('Standard Deviation');
subplot(3,1,3);
loglog(N, sdm);
xlabel('Number of measurements');ylabel('Standard Deviation of Mean');
```

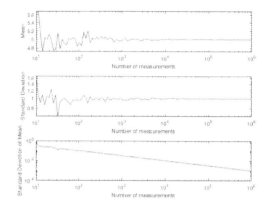

Figure 2.1: Simulating $\overline{y}$, σ, and $\sigma/\sqrt{N}$

2.6   Identifying dubious claims

If you can calculate the best estimate and its uncertainty from repeated measurements, you are already doing pretty well scientifically. You can evaluate if someone else understands significant digits. For example, on TV you might hear, "By 2100 the average global temperature could increase by 2 degrees." There should be a red light that goes off in your head. Could? What are you talking about Fahrenheit or Centigrade? That makes a big difference. How come you end up with this nice number 2? Are you sure it is not 1 or 3 or 0 or even -1? How about 2.1 or 1.9? What is the uncertainty on that prediction? It may be true that the temperature increases is predicted to be 2.00(12)C. Why don't they say something like that?

To be clear, I am not against global warming. I just think the sentence is not precise. You should now understand the difference between a scientific statement backed up my mathematics versus just guessing and estimation. Real scientific statements come with an uncertainty and are not just arbitrary round numbers that sound nice.

2.7 Summary

- Always round to two significant digits of uncertainty.

- Always calculate the uncertainty first.

- Always match the precision of the measurement to the uncertainty.

- Always report the measurement in SI units.

- Mean

$$\overline{y} = \frac{1}{N} \sum_{i=1}^{N} y_i$$

- Variance

$$\sigma^2 = \frac{1}{N} \sum_{i=1}^{N} (y_i - \overline{y})^2$$

- Standard deviation

$$\sigma = \sqrt{\text{Variance}}$$

- Standard deviation of the mean

$$\overline{\sigma} = \frac{\sigma}{\sqrt{N}}$$

Chapter 3

Propagation of Error

3.1 A need for error propagation

3.1.1 Functions of a measurement

Suppose you have made repeated measurements of some angle and determined the uncertainty. If you found the standard deviation of the mean and use radians as units, you may have found something like

$$\theta \pm \overline{\sigma}_\theta = 0.45 \pm 0.12$$

Suppose you want to calculate $\sin \theta$ and its uncertainty. The best way to calculate $\sin \theta$ of course is to use the value you calculated for θ. I am sure you are familiar with that part. Clearly you can't use the same uncertainty you used for the angle.

$$\overline{\sigma}_{\sin \theta} \neq \overline{\sigma}_\theta$$

You have to take some function of the uncertainty and the angle. This is where propagation of error comes in. You have to use an accepted convention to derive the uncertainty for $\sin \theta$. It turns out the formula is

$$\overline{\sigma}_{\sin \theta} = |\cos \theta| \overline{\sigma}_\theta$$

The bars represent absolute value and this is a minor point. You wouldn't want a situation where the error is reported as a negative number. You don't say $x \mp \overline{\sigma}_x$ you always say $x \pm \overline{\sigma}_x$. We will learn how to derive formulas like this in this chapter.

3.1.2 Calculating derived quantities and their uncertainties

A second way error propagation can come into play is if you want to calculate a derived quantity. Suppose you have measured the distance traveled and the elapsed time. Now, you want to calculate the average velocity. Suppose you have

$$d \pm \overline{\sigma}_d = 100.3 \pm 1.2 \, \text{m} \quad t \pm \overline{\sigma}_t = 211.2 \pm 2.2 \, \text{s}$$

Average velocity is a derived quantity because you need to combine two variables to calculate it.

$$v = \frac{d}{t}$$

So you are going to need a new error propagation formula for the ratio of two variables. It turns out that

$$\overline{\sigma}_v = v\sqrt{\frac{\overline{\sigma}_t^2}{t^2} + \frac{\overline{\sigma}_d^2}{d^2}}$$

I am sure by now you can plug in the numbers on a graphing calculator and round to two significant digits to find the uncertainty. Then you could calculate d/t and round that number too. We nailed that calculation in the previous chapter. This chapter is about finding error propagation formulas.

3.1.3   The accepted convention uses calculus

To find these error propagation formulas, you need to use calculus. Some books give several different versions of the error propagation formula. This might seem good if they don't involve calculus and are simple to use for people unsophisticated in math. The problem with this is that different people are using different formulas and end up with different results. If you want to do serious error analysis, look up your formulas in a table in this book or learn calculus and derive them yourself. Everyone can agree only when we follow a convention and derive the same results.

3.2   Differentiation

3.2.1   Ordinary derivatives

I will assume you have had some experience with calculus and finding derivatives. You need to understand ordinary and partial derivatives to do error propagation. Derivatives are functions defined in calculus. The ordinary derivative is defined as

$$f'(x) = \frac{df}{dx} = \lim_{\Delta x \to 0} \frac{f(x + \Delta x) - f(x)}{\Delta x}$$

3.2.2   Rules of differential calculus

If you have an elementary function you can always calculate the derivative if you know enough rules. A software that can calculate derivatives is Mathematica. I have also seen the modern graphing calculators like TI-Nspire CX CAS can take derivatives. You can get Mathematica free with a raspberry pi computer. If you want to master differentiation memorize all these formulas.

$$c' = 0 \qquad x' = 1 \qquad (x^n)' = nx^{n-1} \qquad |x|' = \frac{|x|}{x}$$

$$(cu)' = cu' \qquad (c_1 u + c_2 v)' = c_1 u' + c_2 v'$$

$$(uv)' = u'v + uv' \qquad (1/u)' = -u'/u^2 \qquad (u/v)' = \frac{u'v - uv'}{v^2}$$

$$f(u(x))' = \frac{df}{du}\frac{du}{dx} = f'(u)u'(x)$$

$$(e^x)' = e^x \qquad (a^x)' = a^x \ln a \qquad (\ln x)' = \frac{1}{x} \qquad (\log_a x)' = \frac{1}{x \ln a}$$

$$\frac{d}{dx}\sin x = +\cos x \qquad \frac{d}{dx}\tan x = +\sec^2 x \qquad \frac{d}{dx}\sec x = +\sec x \tan x$$

$$\frac{d}{dx}\cos x = -\sin x \qquad \frac{d}{dx}\cot x = -\csc^2 x \qquad \frac{d}{dx}\csc x = -\csc x \cot x$$

$$\frac{d}{dx}\sinh x = +\cosh x \qquad \frac{d}{dx}\tanh x = +\text{sech}^2 x \qquad \frac{d}{dx}\text{sech}\, x = -\text{sech}\, x \tanh x$$

$$\frac{d}{dx}\cosh x = +\sinh x \qquad \frac{d}{dx}\coth x = -\text{csch}^2 x \qquad \frac{d}{dx}\text{csch}\, x = -\text{csch}\, x \coth x$$

$$\frac{d}{dx}\arcsin x = +\frac{1}{\sqrt{1-x^2}} \qquad \frac{d}{dx}\arctan x = +\frac{1}{1+x^2} \qquad \frac{d}{dx}\text{arcsec}\, x = +\frac{1}{|x|\sqrt{x^2-1}}$$

$$\frac{d}{dx}\arccos x = -\frac{1}{\sqrt{1-x^2}} \qquad \frac{d}{dx}\text{arccot}\, x = -\frac{1}{1+x^2} \qquad \frac{d}{dx}\text{arccsc}\, x = -\frac{1}{|x|\sqrt{x^2-1}}$$

$$\frac{d}{dx}\text{arsinh}\, x = +\frac{1}{\sqrt{x^2+1}} \qquad \frac{d}{dx}\text{artanh}\, x = +\frac{1}{1-x^2} \qquad \frac{d}{dx}\text{arsech}\, x = -\frac{1}{x\sqrt{1-x^2}}$$

$$\frac{d}{dx}\text{arcosh}\, x = +\frac{1}{\sqrt{x^2-1}} \qquad \frac{d}{dx}\text{arcoth}\, x = +\frac{1}{1-x^2} \qquad \frac{d}{dx}\text{arcsch}\, x = -\frac{1}{|x|\sqrt{1+x^2}}$$

3.2.3 Partial derivatives

If you have a function of multiple variables then the differential change in the function uses partial derivatives

$$df(x, y, z) = \frac{\partial f}{\partial x}dx + \frac{\partial f}{\partial y}dy + \frac{\partial f}{\partial z}dz$$

Compare that with a single variable

$$df(x) = f'(x)dx$$

The partial derivative is like the normal derivative except you hold the other variables constant when you take it. Here is the definition

$$\frac{\partial f(x, y)}{\partial x} = \lim_{\Delta x \to 0} \frac{f(x + \Delta x, y) - f(x, y)}{\Delta x}$$

$$\frac{\partial f(x,y)}{\partial y} = \lim_{\Delta y \to 0} \frac{f(x, y + \Delta y) - f(x,y)}{\Delta y}$$

Example 3.1. Let $f(x, y, z) = xy^2 z^3$. Find all the partial derivatives.

Solution 3.1. Recall the power rule $(x^n)' = nx^{n-1}$. There are three possible partial derivatives

$$\frac{\partial f}{\partial x} = y^2 z^3 \quad \frac{\partial f}{\partial y} = 2xyz^3 \quad \frac{\partial f}{\partial z} = 3xy^2 z^2$$

3.3   Error propagation with one variable

3.3.1   The rule for a single variable

Using repeated measurement, one has calculated $\overline{A}$ with an uncertainty $\overline{\sigma}_A$. The single variable propagation of error formula for $\overline{\sigma}_f$ is given by

$$\overline{\sigma}_f = \left| f'(\overline{A}) \right| \overline{\sigma}_A$$

Calculate the uncertainty keeping all your digits and then round to two significant digits. After you have the uncertainty, you can round the measurement $f(\overline{A})$ to match the precision of the uncertainty.

3.3.2   Some examples with a single variable

Example 3.2. Suppose a machinist has constructed a precise cube on a milling machine, and you want to find its volume. You repeatedly measure one of the sides to be

$$s = 1.053 \pm 0.010 \, \text{cm}$$

What is the volume and its uncertainty assuming the length, width, and height are identical?

Solution 3.2. We calculate the volume as $V = s^3$ and wait to round until we know the uncertainty.
$$V = (1.053 \, \text{cm})^3 = 1.16757587 \ldots \, \text{cm}^3$$
To find the uncertainty we use the error propagation formula

$$\overline{\sigma}_V = |V'(s)| \overline{\sigma}_s = 3s^2 \overline{\sigma}_s$$
$$\overline{\sigma}_V = 3(1.053 \, \text{cm})^2 (0.010 \, \text{cm}) = 0.033502 \, \text{cm}^3$$

After rounding we get $V = 1.168(34) \, \text{cm}^3$.

3.3.3 Doing the calculation in MATLAB

Example 3.3. Let $N = 3.2524(35)$ find $f(N) = e^N$ and the uncertainty.

Solution 3.3. We know that $(e^x)' = e^x$. So we easily calculate the uncertainty using

$$\overline{\sigma}_f = |e^N|\overline{\sigma}_N = e^N\overline{\sigma}_N$$

The MATLAB calculation looks like this

```
format long;
N = 3.2524;
sigN = 0.0035;
f = exp(N)
sigf = exp(N)*sigN
>> f = 25.852311068629906
>> sigf = 0.090483088740205
```

We do the rounding and find $f = 25.852(90)$

Example 3.4. What is the uncertainty in $\sin\theta$ when $\theta = 30.0° \pm 2.5°$

Solution 3.4. First we convert the numbers to radians. The error propagation formula is then

$$\overline{\sigma}_f = |\cos\theta|\overline{\sigma}_\theta$$

It is easy to do all these calculations in MATLAB.

```
format long;
t = 30;
sigt = 2.5;
trad = 30 * pi/180;
sigtrad = sigt * pi/180;
df = cos(trad)*sigtrad
f = sin(trad)
>> 0.03778748675488
>> 0.50000000000000
```

So the answer is $\sin\theta = 0.500(38)$.

Example 3.5. Let $A = 0.455$ and $\sigma_A = 0.012$. Calculate with the uncertainty the following quantities: A^2, $\cos(A)$, $\ln(A)$, $\arctan(A)$

Solution 3.5. The derivatives of those functions are respectively $2A$, $-\sin(A)$, $1/A$, and $\frac{1}{1+A^2}$. We can do this all on Matlab fairly quickly

```
A = 0.455;
sigA = 0.012;
A^2
2*A*sigA
cos(A)
abs(-sin(A))*sigA
log(A)
sigA/A
atan(A)
sigA/(1+A^2)
Output
>> A^2
ans =   0.20703
>> 2*A*sigA
ans =   0.010920
>> cos(A)
ans =   0.89826
>> abs(-sin(A))*sigA
ans =   0.0052735
>> log(A)
ans = -0.78746
>> sigA/A
ans =   0.026374
>> atan(A)
ans =   0.42700
>> sigA/(1+A^2)
ans =   0.0099418
```

So the answers are $A^2 = 0.21(11)$, $\cos(A) = 0.8983(53)$, $\ln A = -0.787(26)$, $\arctan(A) = 0.4270(99)$.

Example 3.6. Given $r \pm \overline{\sigma}_r$ find the error in the diameter of a circle, the circumference of a circle, the area of a circle, the surface area of a sphere, the volume of a sphere.

Solution 3.6. To solve these problems recall the power rule $y = r^n$ then $y' = nr^{n-1}$. The diameter of a circle is $2r$ so the uncertainty is $2\overline{\sigma}_r$. The circumference of a circle is $2\pi r$ so the uncertainty is $2\pi\overline{\sigma}_r$. The area of a circle is πr^2 so the uncertainty is $2\pi r\overline{\sigma}_r$. The surface area of a circle is $4\pi r^2$ so the uncertainty is $8\pi r\overline{\sigma}_r$. The volume of a sphere is $\frac{4}{3}\pi r^3$ so the uncertainty is $4\pi r^2\overline{\sigma}_r$.

3.4 Error propagation for multiple independent variables

3.4.1 The rule for multiple independent variables

Often we want to calculate the uncertainty of a derived quantity from multiple measurements For example, if we wanted to calculate AB we would first measure A and its uncertainty then measure B and its uncertainty. To find the error in AB we will need to apply a conventional formula of error propagation. There is a significant simplification if the simultaneous measurements of A and B are independent. For example, the charge and mass of an electron are independent.

The error propagation formula for multiple independent variables labeled $a_1, a_2, \ldots, a_M$ is given by

$$\overline{\sigma}_f^2 = \sum_{j=1}^{M} \left(\frac{\partial f}{\partial a_j} \right)^2 \overline{\sigma}_{a_j}^2$$

We will often use the notation that $a_1 = A \quad a_2 = B$ and so on to simplify the formulas. Propagation of error for two independent variables is given by

$$\overline{\sigma}_f^2 = \left(\frac{\partial f}{\partial A} \right)^2 \overline{\sigma}_A^2 + \left(\frac{\partial f}{\partial B} \right)^2 \overline{\sigma}_B^2$$

3.4.2 Common formulas sums and differences

Example 3.7. Show that $\overline{\sigma}_{A \pm B} = \sqrt{\overline{\sigma}_A^2 + \overline{\sigma}_B^2}$

Solution 3.7. Evaluating the partial derivatives

$$\frac{\partial(A \pm B)}{\partial A} = 1 \qquad \frac{\partial(A \pm B)}{\partial B} = \pm 1$$

Applying the error propagation formula we have

$$\overline{\sigma}_{A \pm B} = \sqrt{(1)^2 \overline{\sigma}_A^2 + (\pm 1)^2 \overline{\sigma}_B^2} = \sqrt{\overline{\sigma}_A^2 + \overline{\sigma}_B^2}$$

3.4.3 Common formulas products and quotients

Example 3.8. If $f = AB$, then show that

$$\overline{\sigma}_{AB} = |AB| \sqrt{\frac{\overline{\sigma}_A^2}{A^2} + \frac{\overline{\sigma}_B^2}{B^2}}$$

Solution 3.8. Evaluating the partial derivatives

$$\frac{\partial(AB)}{\partial A} = B \qquad \frac{\partial(AB)}{\partial B} = A$$

Applying the error propagation formula we have

$$\overline{\sigma}_{AB} = \sqrt{B^2\overline{\sigma}_A^2 + A^2\overline{\sigma}_B^2}$$

Divide both sides by $|AB|$.

$$\frac{\overline{\sigma}_{AB}}{|AB|} = \frac{1}{|AB|}\sqrt{B^2\overline{\sigma}_A^2 + A^2\overline{\sigma}_B^2} = \sqrt{\frac{\overline{\sigma}_A^2}{A^2} + \frac{\overline{\sigma}_B^2}{B^2}}$$

Example 3.9. If $f = A/B$, then show that

$$\overline{\sigma}_{AB} = |A/B|\sqrt{\frac{\overline{\sigma}_A^2}{A^2} + \frac{\overline{\sigma}_B^2}{B^2}}$$

Solution 3.9. Evaluating the partial derivatives

$$\frac{\partial(A/B)}{\partial A} = 1/B \qquad \frac{\partial(A/B)}{\partial B} = -A/B^2$$

Applying the error propagation formula we have

$$\overline{\sigma}_{A/B} = \sqrt{(1/B)^2\overline{\sigma}_A^2 + (A^2/B^4)\overline{\sigma}_B^2}$$

Divide both sides by $|A/B|$.

$$\frac{\overline{\sigma}_{A/B}}{|A/B|} = \frac{1}{|A/B|}\sqrt{(1/B)^2\overline{\sigma}_A^2 + (A^2/B^4)\overline{\sigma}_B^2} = \sqrt{\frac{\overline{\sigma}_A^2}{A^2} + \frac{\overline{\sigma}_B^2}{B^2}}$$

3.4.4   Other common error formulas

- $f = AB/C \ldots$ then

$$\frac{\overline{\sigma}_f}{|f|} = \sqrt{\frac{\overline{\sigma}_A^2}{A^2} + \frac{\overline{\sigma}_B^2}{B^2} + \frac{\overline{\sigma}_C^2}{C^2} + \ldots}$$

- $f = A^a B^B/C^c \ldots$ then

$$\frac{\overline{\sigma}_f}{|f|} = \sqrt{\frac{a^2\overline{\sigma}_A^2}{A^2} + \frac{b^2\overline{\sigma}_B^2}{B^2} + \frac{c^2\overline{\sigma}_C^2}{C^2} + \ldots}$$

Many formulas in physics are of the last form. For example, $F = GmM/r^2$ or $\omega = \sqrt{k/m}$.

3.4.5 MATLAB examples

Example 3.10. Find the perimeter of a rectangle of $l = 1.25(22)$ and $w = 4.44(33)$

Solution 3.10. The perimeter is equal to $P = 2(l + w)$ We can find the perimeter by plugging in the values $P = 2(1.25 + 4.44) = 11.3800\dots$ We can find the uncertainty by applying the propagation of error formula

$$\overline{\sigma}_P^2 = \left(\frac{\partial P}{\partial l}\right)^2 \overline{\sigma}_l^2 + \left(\frac{\partial P}{\partial w}\right)^2 \overline{\sigma}_w^2 = 4\overline{\sigma}_l^2 + 4\overline{\sigma}_w^2$$

Here is the MATLAB code

```
l = 1.25;
sigl = 0.22;
w = 4.44;
sigw = 0.33;
P = 2*(l + w);
sigP = sqrt( (4)*sigl^2 + 4*sigw^2)
>> P = 11.380
>> sigP = 0.79322
```

So the final perimeter of the rectangle is $P = 11.38(79)$ after rounding.

Example 3.11. Find the value of the Stefan Boltzmann constant and its uncertainty

Solution 3.11. The Stefan Boltzmann constant is

$$\sigma = \frac{\pi^2 k_B^4}{60\hbar^3 c^2}$$

It's funny that the notation for the Stefan Boltzmann constant is already σ, don't get confused. Using the common propagation of error formulas

$$\overline{\sigma}_\sigma = \sigma \sqrt{4^2 \frac{\overline{\sigma}_{k_B}^2}{k_B^2} + 3^2 \frac{\overline{\sigma}_\hbar^2}{\hbar^2}}$$

Since c is defined it has no uncertainty. Also constant factors like π or $\sqrt{2}$ have no uncertainty. A quick google for "NIST hbar", "NIST c", and "NIST Boltzmann's constant" gives the necessary data.

```
format long;
hb   = 1.054571800e-34;
shb = 0.000000013e-34;
c = 299792458;
```

```
kb    = 1.38064852e-23;
skb = 0.00000079e-23;


SB   = pi^2/60 * kb^4/(c^2*hb^3)
sigSB = SB*(16*skb^2/kb^2+9*shb^2/hb^2)^(0.5)
>> 5.670366818327269e-08
>> 1.297991325923970e-13
```

The uncertainty is 0.000013×10^{-8} J m^{-2}s^{-1}K^{-4}. So the final result is

$$\sigma = 5.670367(13) \times 10^{-8} \text{Jm}^{-2}\text{s}^{-1}\text{K}^{-4}$$

This calculation agrees with the NIST value.

Example 3.12. Find the value of the Planck mass and its uncertainty

Solution 3.12. The Planck mass is defined as the dimensionless combination of c, $\hbar$, and G that gives units of mass.

$$m_P = \sqrt{\frac{\hbar c}{G}}$$

The propagation of error formula is then

$$\overline{\sigma}_{m_P} = m_P \sqrt{(1/2)^2 \frac{\overline{\sigma}_\hbar^2}{\hbar^2} + (-1/2)^2 \frac{\overline{\sigma}_G^2}{G^2}}$$

Doing the MATLAB calculation we have

```
hb    = 1.054571800e-34;
shb = 0.000000013e-34;
c = 299792458;
G = 6.67408e-11
sG = 0.00031e-11
mP = (hb*c/G)^(0.5)
smP = mP*(1/4*shb^2/hb^2+1/4*sG^2/G^2)^(0.5)
>> 2.176470195490643e-08
>> 5.054672586812261e-13
```

$$m_P = 2.176470(51) \times 10^{-8}\text{kg}$$

This calculation agrees with the NIST value.

Example 3.13. A relativistic particle has an energy $E = \sqrt{(pc)^2 + (mc^2)^2}$. If one is given $p \pm \overline{\sigma}_p$ and $m \pm \overline{\sigma}_m$ find the uncertainty in the energy.

Solution 3.13. We just need to find the partial derivatives.

$$\frac{\partial E}{\partial p} = (1/2)\frac{2pc^2}{\sqrt{(pc)^2 + (mc^2)^2}} = \frac{pc}{E}c$$

$$\frac{\partial E}{\partial m} = (1/2)\frac{2mc^4}{\sqrt{(pc)^2 + (mc^2)^2}} = \frac{mc^2}{E}c^2$$

$$\overline{\sigma}_E = \sqrt{\frac{(pc)^2}{E^2}\overline{\sigma}_p^2 c^2 + \frac{(mc^2)^2}{E^2}\overline{\sigma}_m^2 c^4}$$

One can see the units work out right

Example 3.14. Consider Snell's law $n_1 \sin(\theta_i) = n_2 \sin(\theta_r)$ If $n_1 = 1.234 \pm 0.022$ and $n_2 = 1.344 \pm 0.012$ then if the angle of incidence is 15.02 ± 0.13 degrees then what is the angle of refraction.

Solution 3.14. First convert the angle into radians by multiplying by $\pi/180$. The angle of incidence is $0.2622(23)$ radians. It is easier to solve for $\sin(\theta_r)$.

$$\sin(\theta_r) = (n_1/n_2)\sin(\theta_i)$$

Using the formula for dividing to variables we find $n_1/n_2 = 0.918(18)$ Using the formula for $\sin\theta$ we find

$$\sin(\theta_r) = \sin(0.2622) \pm \cos(0.2622)(0.0023) = 0.2592(22)$$

Now $\sin\theta_i \times (n_1/n_2)$ is given by $0.2379(51)$. Now we have to find the arcsine.

$$\overline{\theta}_r = \arcsin\left(0.2379(51)\right)$$

This is just a single variable formula so we can take the derivative and find.

$$\overline{\sigma}_{\theta_r} = \frac{1}{\sqrt{1 - \theta_r^2}}\overline{\sigma}_{\theta_r} = (1/\sqrt{1 - 0.2379^2})(0.0051) = 0.0053$$

So the angle of refraction is $0.2402(53)$ radians 0r $13.76(30)$ degrees. As you can see there is a lot of work in doing propagation of error to get correct answers.

3.5 Derivation of the error propagation formulas

3.5.1 The main formula

Consider setting an experiment to some setting x and then measuring two experimental variables A and B from a series of N measurements each. We can then refer to their means and standard deviations.

$$\overline{A} = \frac{1}{N}\sum_{j=1}^{N} A_j$$

$$\sigma_A^2 = \frac{1}{N}\sum_{j=1}^{N} (A_j - \overline{A})^2$$

$$\overline{B} = \frac{1}{N}\sum_{j=1}^{N} B_j$$

$$\sigma_B^2 = \frac{1}{N}\sum_{j=1}^{N} (B_j - \overline{B})^2$$

We could also find the standard deviations of the mean.

$$\overline{\sigma}_A = \sigma_A/\sqrt{N} \qquad \overline{\sigma}_B = \sigma_B/\sqrt{N}$$

Now we want to consider a model $f(A, B)$. The model could be something like $f = A + B$ and we want to find the uncertainty in f as a functions of the above information. Then we would report $A + B \pm \overline{\sigma}_{A+B}$. The conventional way to treat this problem is to consider a Taylor series. We have to measure each A and B a total of N times so I will index them by j.

$$f(A_j, B_j) \approx f(\overline{A}, \overline{B}) + \frac{\partial f}{\partial A}(A_j - \overline{A}) + \frac{\partial f}{\partial B}(B_j - \overline{B}) + \dots$$

The partial derivatives are evaluated at $(\overline{A}, \overline{B})$. Now we move the term $f(\overline{A}, \overline{B})$ to the left hand side, square both sides, divide both sides by N, and sum over j from 1 to N.

$$\sigma_f^2 = \left(\frac{\partial f}{\partial A}\right)^2 \sigma_A^2 + \left(\frac{\partial f}{\partial B}\right)^2 \sigma_B^2 + 2\left(\frac{\partial f}{\partial A \partial B}\right)\frac{1}{N}\sum_{i=1}^{N}(A_j - \overline{A})(B_j - \overline{B})$$

3.5.2   Correlated variables and covariance

We get cross terms that involve the covariance.

$$\sigma_f^2 = \left(\frac{\partial f}{\partial A}\right)^2 \sigma_A^2 + \left(\frac{\partial f}{\partial B}\right)^2 \sigma_B^2 + 2\left(\frac{\partial f}{\partial A \partial B}\right)\text{cov}(A, B)$$

It is more natural to write the covariance in terms of Pearson's correlation coefficient.

$$\rho_{AB} = \frac{\text{cov}(A, B)}{\sigma_A \sigma_B} = \frac{\frac{1}{N}\sum_{j=1}^{N}(A_j - \overline{A})(B_j - \overline{B})}{\sigma_A \sigma_B}$$

Notice that $\rho_{AA} = \rho_{BB} = 1$. We are usually interested in the standard deviation of the mean of the variables. Since we took N measurements each for A and B, we can divide both sides of the equation by N and get.

$$\overline{\sigma}_f^2 = \left(\frac{\partial f}{\partial A}\right)^2 \overline{\sigma}_A^2 + \left(\frac{\partial f}{\partial B}\right)^2 \overline{\sigma}_B^2 + 2\left(\frac{\partial f}{\partial A \partial B}\right)\overline{\sigma}_A \overline{\sigma}_B \rho_{AB}$$

The generalization to M variables is straightforward now. We use $\mathbf{a} = (a_1, a_2, \ldots, a_M)$.

$$\overline{\sigma}_f^2 = \sum_{j=1}^{M} \sum_{k=1}^{M} \left(\frac{\partial f}{\partial a_j} \right) \left(\frac{\partial f}{\partial a_k} \right) \overline{\sigma}_{a_j} \overline{\sigma}_{a_k} \rho_{jk}$$

One can even write this expression in matrix notation.

$$\overline{\sigma}_f^2 = \mathbf{J}^T \mathbf{V} \mathbf{J}$$

$$V_{jk} = \overline{\sigma}_j \overline{\sigma}_k \rho_{jk}$$

$$J_j = \frac{\partial f}{\partial a_j}$$

3.5.3 Error propagation with covariance

Example 3.15. Find the error propagation formula for $aA + bB$ if the A and B are correlated and a and b are constants.

Solution 3.15. We have the error propagation formula

$$\overline{\sigma}_f^2 = \sum_{j=1}^{M} \sum_{k=1}^{M} \left(\frac{\partial f}{\partial a_j} \right) \left(\frac{\partial f}{\partial a_k} \right) \overline{\sigma}_{a_j} \overline{\sigma}_{a_k} \rho_{jk}$$

Taking the derivatives,

$$\overline{\sigma}_f^2 = a^2 \overline{\sigma}_A^2 + b^2 \overline{\sigma}_B^2 + 2ab\overline{\sigma}_A \overline{\sigma}_B \rho_{AB}$$

Example 3.16. Find the error propagation formula for $A - B$ if the variables are correlated.

Solution 3.16. Similar to the previous problem

$$\overline{\sigma}_f^2 = a^2 \overline{\sigma}_A^2 + b^2 \overline{\sigma}_B^2 - 2ab\overline{\sigma}_A \overline{\sigma}_B \rho_{AB}$$

Example 3.17. If you look at the formula for covariance for $A - B$ you see that you have to subtract a term for the correlations. There is some chance that the covariance term will cancel out the other terms and you get zero. What do you reason you should do in this situation? Can the error actually be zero?

Solution 3.17. The error can't be zero. This means that the second order Taylor expansion we took was not accurate enough. You would have to go to higher orders to calculate the actual error.

Example 3.18. Find the error propagation formula for AB if the variables are correlated.

Solution 3.18.

$$\overline{\sigma}_f^2 = B^2\overline{\sigma}_A^2 + A^2\overline{\sigma}_B^2 + 2AB(\overline{\sigma}_A\overline{\sigma}_B)\rho_{AB}$$

$$\overline{\sigma}_f^2/(AB)^2 = \frac{\overline{\sigma}_A^2}{A^2} + \frac{\overline{\sigma}_B^2}{B^2} + \frac{2}{AB}(\overline{\sigma}_A\overline{\sigma}_B)\rho_{AB}$$

Example 3.19. Find the error propagation formula for A/B if the variables are correlated.

Solution 3.19.

$$\overline{\sigma}_f^2 = (1/B)^2\overline{\sigma}_A^2 + (A^2/B^4)\overline{\sigma}_B^2 + 2(1/B)(-A/B^2)(\overline{\sigma}_A\overline{\sigma}_B)\rho_{AB}$$

$$\overline{\sigma}_f^2/(A/B)^2 = \frac{\overline{\sigma}_A^2}{A^2} + \frac{\overline{\sigma}_B^2}{B^2} - \frac{2}{AB}(\overline{\sigma}_A\overline{\sigma}_B)\rho_{AB}$$

3.5.4   Reduction of the error propagation formula

An important approximation, or rather constraint is that the covariances are equal to zero. For example, the measurement of Planck's constant does not dependent not depend on the simultaneous measurement of the charge of an electron. When you work with real data of independent variables and actually calculate the covariance, you will get some small number just due to random error. If you choose to neglect it, then we arrive at a simpler equation called the independent variable propagation of error formula.

$$\overline{\sigma}_f^2 = \sum_{j=1}^{M} \left(\frac{\partial f}{\partial a_j}\right)^2 \overline{\sigma}_{a_j}^2$$

One can also see how this formula reduces to

$$\overline{\sigma}_f = |f'(\overline{A})|\overline{\sigma}_A$$

in the single variable case.

3.6   Error propagation step by step

3.6.1   Doing step by step error propagation

Suppose we have a formula

$$f(A, B, C, D) = \frac{A + B}{C + D}$$

and we want to find the error propagation formula. It is convenient to do it step by step when you know some basic formulas of error propagation. We can let $Y = A + B$ and $Z = C + D$ then find the uncertainties in Y and Z first.

$$\overline{\sigma}_Y = \sqrt{\overline{\sigma}_A^2 + \overline{\sigma}_B^2} \qquad \overline{\sigma}_Z = \sqrt{\overline{\sigma}_C^2 + \overline{\sigma}_D^2}$$

We can then calculate the uncertainty for Y/Z which would be

$$\frac{\overline{\sigma}_f}{|f|} = \sqrt{\frac{\overline{\sigma}_Y^2}{Y^2} + \frac{\overline{\sigma}_Z^2}{Z^2}}$$

That is a way to calculate the formula step by step. You can check for yourself that you get the same result as if you did the calculation using the direct formula.

3.6.2 When you can't use step by step

Some formulas do now allow you to do step by step error propagation. If the formula was

$$f = \frac{A - B}{A + B}$$

then we couldn't do it step by step the same way because the numerator and the denominator contain repeating variables. Partial derivatives are taken for the whole function top and bottom so don't make the mistake of misapply step by step error propagation.

3.6.3 Step by step examples

Example 3.20. Find the uncertainty in $E = (A + B)(C + D)$

Solution 3.20. We can do step by step propagation. The first term has an uncertainty $\overline{\sigma}_1 = \sqrt{\overline{\sigma}_A^2 + \overline{\sigma}_B^2}$. The second term has an uncertainty $\overline{\sigma}_2 = \sqrt{\overline{\sigma}_C^2 + \overline{\sigma}_D^2}$. The step by step uncertainty is

$$\overline{\sigma}_E = E\sqrt{\overline{\sigma}_1^2/(A+B)^2 + \overline{\sigma}_2^2/(C+D)^2}$$

Example 3.21. Find the uncertainty in

$$E = \frac{A^2 + B^2}{C^2 + D^2}$$

Solution 3.21. The uncertainty of the numerator is

$$\sigma_1 = \sqrt{(2A)^2\sigma_A^2 + (2B)^2\sigma_B^2}$$

The uncertainty of the denominator is also

$$\sigma_2 = \sqrt{(2C)^2\sigma_C^2 + (2D)^2\sigma_D^2}$$

So the total uncertainty is

$$\frac{\sigma_E^2}{E^2} = \frac{\sigma_1^2}{(A^2+B^2)^2} + \frac{\sigma_2^2}{C^2+D^2}$$

Example 3.22. Find

$$(0.92(33) + 1.29(27))(1.92(51) - 3.24(11))$$

Solution 3.22. We can calculate the two factors first.

$$[0.92(33) + 1.29(27)] = 2.21(43) \qquad [1.92(51) - 3.24(11)] = -1.32(52)$$

Now we can multiply the two numbers

$$A = 2.21(43) \times B = -1.32(52)$$

$$\sigma_{AB} = \sqrt{\sigma_A^2 B^2 + \sigma_B^2 A^2}$$

$$AB = -2.9(13)$$

Example 3.23. You have two resistors with measured resistances $R_1 = 101.2 \pm 2.3$ Ohms and $R_2 = 204.1 \pm 3.7$ Ohms. Find the equivalent resistance in series and in parallel.

Solution 3.23. The equivalent resistance in series is $R_{eq} = R_1 + R_2$. The equivalent resistance in parallel is $R_{eq} = \dfrac{R_1 R_2}{R_1 + R_2}$. The equivalent resistance is just then

$$R_{series} = (101.2 + 204.1 \pm \sqrt{2.3^2 + 3.7^2})\Omega = (305.3 \pm 4.4)\Omega$$

Here is a good example where you don't want to try step by step error propagation. Since

$$\frac{\partial}{\partial x}\left(\frac{xy}{x+y}\right) = \frac{y^2}{(x+y)^2}$$

The equivalent resistance uncertainty in parallel is just

$$\overline{\sigma}_p = \sqrt{\frac{\overline{\sigma}_1^2 R_2^4}{(R_1+R_2)^4} + \frac{\overline{\sigma}_2^2 R_1^4}{(R_1+R_2)^4}}$$

This is the MATLAB code

```
R1  = 101.2
R2 = 204.1
sig1 = 2.3
sig2 = 3.7
sigRp = sqrt(sig1^2 * R2^4/(R1 + R2)^4 + sig1^2 * R1^4/(R1 + R2)^4)
Req = R1*R2/(R1 + R2)
```

So the equivalent resistance is $67.7(1.1)\,\Omega$.

3.7 Experimental design to reduce error

3.7.1 Dominant error

The error propagation formula tells you where you should focus your efforts if you want to reduce the uncertainty in a derived quantity. Usually one term will be larger than another and this is called the dominant error. Suppose you are calculating g from experiments with a pendulum with small oscillations.

$$\omega^2 = g/l \qquad g = 4\pi^2 \frac{l}{T^2}$$

If the fractional uncertainty in the length and period are both ten percent, then how should you proceed to improve the experimental determination of g? The error propagation formula for this situation. We can easily write down the error propagation formula

$$\frac{\overline{\sigma}_g}{g} = \sqrt{\frac{\overline{\sigma}_l^2}{l^2} + 4\frac{\overline{\sigma}_T^2}{T^2}}$$

If the fractional uncertainty in the length is 10 percent and the fractional uncertainty in the period is 10 percent then the expression in the square root is

$$\sqrt{0.1^2 + 4 \times 0.1^2}$$

Clearly we should take more measurements of T to reduce it because it is the dominant error term.

3.7.2 The square root counting rule, reducing dominant error

If you want to reduce the standard deviation of the mean you should just repeat more measurements. While the standard deviation is not reduced by taking more measurements the standard deviation of the mean is reduced by a factor of $\sqrt{N}$ where N is the total number of measurements. This is a handy rule to keep in mind

Example 3.24. An experimenter measures $N = 1023 \pm 55$ with 20 measurements. How many more measurements would be expected to reduce the uncertainty to one part in 100? One part in a 1000?

Solution 3.24. The uncertainty starts off to be $55/1023$ or 1 part in 18.6. So $18.6\sqrt{N} = 100$ or $N = 29$. So you need to do 9 more measurements. Similarly

$$18.6\sqrt{N} = 1000 \rightarrow N = 2891$$

So you need to do about 2871 more measurements to get 1 in a thousand. This is not an exact thing but a general rule of thumb. It should get you in the right range to think about how you plan out your experiment and how much uncertainty you want in the end.

Example 3.25. Suppose it costs $100 to take an expensive fluorescence measurement. A student takes a trial measurement and finds 1452 ± 87 photons. Suppose the student has a budget of $2500. He figures that if he buys a more stable laser for $1000 dollars he can reduce his standard deviation by a factor of 2 on the remaining measurements. Then he would have $1500 left to take 15 more measurements. To get the best results on his budget, should the student get the new laser or just continue the measurements on his current setup?

Solution 3.25. Let's see what can be accomplished by buying the laser. If you spend a thousand dollars on the laser you can reduce the uncertainty to about 43. After 15 measurements and neglecting the first measurement, he will get a standard deviation of the mean of $43/\sqrt{15} = 11$ photons. If he just did the 25 measurements more he would get an uncertainty of $87/\sqrt{26} = 17$ photons. He can therefore justify buying the laser to get a better result.

3.8   Summary

- $\overline{\sigma}_f = |f'(A)|\overline{\sigma}_A$ single variable propagation of error

- $\overline{\sigma}_f^2 = \sum_{j=1}^{M} \left(\frac{\partial f}{\partial a_j}\right)^2 \overline{\sigma}_{a_j}^2$ multiple independent variables propagation of error.

- $f = A + B$ then $\overline{\sigma}_f = \sqrt{\overline{\sigma}_A^2 + \overline{\sigma}_B^2}$

- $f = A - B$ then $\overline{\sigma}_f = \sqrt{\overline{\sigma}_A^2 + \overline{\sigma}_B^2}$

- $f = AB$ then $\dfrac{\overline{\sigma}_f}{|f|} = \sqrt{\dfrac{\overline{\sigma}_A^2}{A^2} + \dfrac{\overline{\sigma}_B^2}{B^2}}$

- $f = A/B$ then $\dfrac{\overline{\sigma}_f}{|f|} = \sqrt{\dfrac{\overline{\sigma}_A^2}{A^2} + \dfrac{\overline{\sigma}_B^2}{B^2}}$

- $f = AB/C \ldots$ then $\dfrac{\overline{\sigma}_f}{|f|} = \sqrt{\dfrac{\overline{\sigma}_A^2}{A^2} + \dfrac{\overline{\sigma}_B^2}{B^2} + \dfrac{\overline{\sigma}_C^2}{C^2} + \ldots}$

- $f = A^a B^B / C^c \ldots$ then

$$\frac{\overline{\sigma}_f}{|f|} = \sqrt{\frac{a^2\overline{\sigma}_A^2}{A^2} + \frac{b^2\overline{\sigma}_B^2}{B^2} + \frac{c^2\overline{\sigma}_C^2}{C^2} + \cdots}$$

- General propagation of error

$$\overline{\sigma}_f^2 = \sum_{j=1}^{M} \sum_{k=1}^{M} \left(\frac{\partial f}{\partial a_j}\right)\left(\frac{\partial f}{\partial a_k}\right) \overline{\sigma}_{a_j}\overline{\sigma}_{a_k}\rho_{jk}$$

- $f = A + B,$ $\overline{\sigma}_f^2 = a^2\overline{\sigma}_A^2 + b^2\overline{\sigma}_B^2 + 2ab\overline{\sigma}_A\overline{\sigma}_B\rho_{AB}$

- $f = A - B,$ $\overline{\sigma}_f^2 = a^2\overline{\sigma}_A^2 + b^2\overline{\sigma}_B^2 - 2ab\overline{\sigma}_A\overline{\sigma}_B\rho_{AB}$

- $f = AB,$

$$\overline{\sigma}_f^2 / (A/B)^2 = \frac{\overline{\sigma}_A^2}{A^2} + \frac{\overline{\sigma}_B^2}{B^2} + \frac{2}{AB}(\overline{\sigma}_A\overline{\sigma}_B)\rho_{AB}$$

- $f = A/B,$

$$\overline{\sigma}_f^2 / (A/B)^2 = \frac{\overline{\sigma}_A^2}{A^2} + \frac{\overline{\sigma}_B^2}{B^2} - \frac{2}{AB}(\overline{\sigma}_A\overline{\sigma}_B)\rho_{AB}$$

Chapter 4

Linear Least Squares

4.1   The use of curve fitting

4.1.1   How to establish a theoretical model

If we have spent some energy, time, and resources to take repeated measurements, we can reduce those measurements to a dataset $\{x_i, \overline{y}_i, \overline{\sigma}_i\}$. We would then like to compare our dataset with a theoretical model. The model might predict a relationship that can be observed in data. The model might contain some unknown parameters which can be extracted from the dataset. We use a curve fit to find those parameters and their uncertainties.

In general, we consider a model with one independent variable x at a time and M unknown parameters or parameters that can be determined by the data.

$$y = f(x; a_1, a_2, \ldots, a_M)$$

By choosing different parameters we can evaluate how well our model has the potential to match the data.

Usually one cannot guess the best parameters because the space is very large. If each $\overline{y}_i$ is similar to $f(x_i, \mathbf{a})$ then we see that the model is good. With our knowledge from calculus we might want to set up an error function that we try to minimize.

$$E = \sum_{i=1}^{N} (\overline{y}_i - f(x_i, \mathbf{a})) = \sum_{i=1}^{N} r_i$$

We call r_i the residuals. This doesn't turn out to be very good because sometimes the residuals are positive and sometimes they are negative so they average out to zero. One idea is to take the absolute value of the residuals in our error function.

$$E = \sum_{i=1}^{N} |\overline{y}_i - f(x_i, \mathbf{a})| = \sum_{i=1}^{N} |r_i|$$

You can work with this numerically but you can't get nice analytical results because the absolute value is pathological in calculus.

One way around this is to square the residuals.

$$E = \sum_{i=1}^{N} (\overline{y}_i - f(x_i, \mathbf{a}))^2$$

This is easy to work with in calculus and gives some nice analytical results, but may seem arbitrary. One can justify it with some probability arguments that we can consider later.

To establish the model, you vary one experimental parameter at a time over some range. Then you do the curve fit and show that the error is small for the theoretical model. You do this for each variable, each time displaying and fitting the dependence for each variable. Once you have exhausted all the experimental variables you have provided evidence for your theory. People now have a chance to replicate and reproduce what you have done and then it becomes even more believable.

4.1.2   The range of validity

A good curve fit will allow us to interpolate the data over the measured range. Interpolation means we can estimate the expected value of in between points from the neighboring points. Extrapolation is using the model to predict values outside the measured range. One should do this with some caution because the model might break down or change in extreme limits. You only know something is true in the range in which you measure it. For example, at high velocities near the speed of light Newtonian mechanics breaks down and you need special relativity. The previous measurements are still valid, just that you can't extrapolate them indefinitely.

4.1.3   Determining model parameters and uncertainties

To find the M parameters you are going to need at least $N \geq M$ points. It is better when $N >> M$. The model should be as simple as possible, the fewer M the better. Curve fits with too many parameters are taken with a grain of salt. There is usually some combination of parameters that works when there are many fitting parameters. There may be large space of similar parameters that work equally well and you could have a hard time finding the best ones. This is why it is important to first consider models that have exact results for parameters and uncertainties. We establish a convention. Then there is only one way to fit an analytic model, and everyone can agree. When I say analytic, you still use a computer to get the results, but the computer won't fail because of rounding errors or something like that. An analytic model, can be expressed in terms of algebra and solved by the calculus of optimization.

4.2   The least squares method

4.2.1   Optimisation in calculus

We can now see how curve fitting can be thought of as an optimization problem where you minimize some error function. Let's review some concepts from calculus that allow us to perform optimization now.

Suppose we have a function $f(x)$ and want to find the minimum value of the function over some range $[a, b]$. We know that local minima are critical points where $f'(x_0) = 0$ and the curvature is upwards $f''(x_0) > 0$.

Example 4.1. Find the critical points of $f(x) = A + Bx + Cx^2$. What condition is required for a global minimum.

Solution 4.1. We take the derivative

$$f'(x) = B + 2Cx_0 = 0$$
$$x_0 = -B/2C$$

For x_0 to be a minimum, we require $f''(x_0) = 2C > 0$ or $C > 0$

4.2.2 Linear models

The most basic model you can do a curve fit on is a straight line. We will consider three models. The first is $y = A + Bx$ this is the most general straight line. The second is a straight line through the origin $y = Bx$. The third is a horizontal line $y = A$. This is also the same as averaging together the measurements and is known as the weighted mean. It is weighted because the experiments typically have different uncertainties and you put more preference towards the measurements with smaller uncertainties.

Linear model are more general. Basically any function behind the A and B or C is possible as long as they don't depend on the fit parameters. The parabola $y = A + Bx + Cx^2$ is a linear model also, which sounds kind of weird. You will get the hang of it once we start solving some models and learn what you can do analytically and what you cannot. Sometimes a model is nonlinear, but you can transform it to a linear one. The good news is that if everyone uses the least squares convention everyone will get the same results for the fit parameters and uncertainties for linear models. It is definitely a worthwhile endeavor and something every scientist should learn. You want to know what is under the hood of your statistics program that is doing all these calculations automatically. You should be able to at least check the software to see if you get the same results in some simple situations.

4.2.3 Optimisation of χ^2

The error function E we considered before can use one more tweak. We want to weight the points that we know with smaller uncertainties to count more in the optimization and the points with large error bars to count less. To do this, instead of E we define χ^2.

$$\chi^2 = \sum_{i=1}^{N} \frac{(\overline{y}_i - f(x_i, \mathbf{a}))^2}{\overline{\sigma}_i^2}$$

The weighting is

$$w_i = \frac{1}{\overline{\sigma}_i^2}$$

This makes χ^2 dimensionless. We use the notation that $\overline{\mathbf{a}}$ give the smallest χ^2. χ^2 is always the starting point for curve fitting with the least squares convention.

4.3   FItting a straight line

4.3.1   Fitting $y = A + Bx$

The model for the general straight line is $f(x_i; A, B) = A + Bx$. We recognize B as the slope, and A the y intercept. We can apply the definition of χ^2 and try to minimize it.

$$\chi^2 = \sum_{i=1}^{N} \frac{(\overline{y}_i - A - Bx_i)^2}{\overline{\sigma}_i^2}$$

Expanding the 9 terms gives

$$\chi^2 = A^2 S_w + 2AB S_{wx} + B^2 S_{wxx} - 2A S_{wy} - 2B S_{wxy} + S_{wyy}$$

We use the notation that S_{wuv} is given by

$$S_{wuv} = \sum_{i=1}^{N} w_i u_i v_i \qquad \text{where } w_i = \frac{1}{\overline{\sigma}_i^2}$$

Since there are two unknown parameters we require the condition that

$$\frac{\partial}{\partial A}\chi^2 = 0 \qquad \frac{\partial}{\partial B}\chi^2 = 0$$

This gives two equation which can be solved for $\overline{A}$ and $\overline{B}$.

$$\overline{A} S_w + \overline{B} S_{wx} = S_{wy}$$
$$\overline{A} S_{wx} + \overline{B} S_{wxx} = S_{wxy}$$

We can solve it with Cramer's rule.

$$\overline{A} = \frac{\begin{vmatrix} S_{wy} & S_{wx} \\ S_{wxy} & S_{wxx} \end{vmatrix}}{\begin{vmatrix} S_w & S_{wx} \\ S_{wx} & S_{wxx} \end{vmatrix}} = \frac{S_{wy}S_{wxx} - S_{wx}S_{wxy}}{S_w S_{wxx} - S_{wx}^2}$$

$$\overline{B} = \frac{\begin{vmatrix} S_w & S_{wy} \\ S_{wx} & S_{wxy} \end{vmatrix}}{\begin{vmatrix} S_w & S_{wx} \\ S_{wx} & S_{wxx} \end{vmatrix}} = \frac{S_w S_{wxy} - S_{wx}S_{wy}}{S_w S_{wxx} - S_{wx}^2}$$

A common definition is $\Delta = S_w S_{wxx} - S_{wx}^2$.

4.3.2   Uncertainty in the fit parameters

Now we find the uncertainties in A and B by using the propagation of error formula.

$$\overline{\sigma}_A^2 = \sum_{i=1}^{N} \left(\frac{\partial A}{\partial \overline{y}_i}\right)^2 \overline{\sigma}_i^2$$

$$\overline{\sigma}_A^2 = (1/\Delta^2) \sum_{i=1}^{N} (w_i S_{wxx} - S_{wx} w_i x_i)^2 (1/w_i) = \frac{S_{wxx}}{\Delta}$$

$$\overline{\sigma}_B^2 = \sum_{i=1}^{N} \left(\frac{\partial B}{\partial \overline{y}_i}\right)^2 \overline{\sigma}_i^2$$

$$\overline{\sigma}_B^2 = (1/\Delta^2) \sum_{i=1}^{N} (S_w w_i x_i - S_{wx} w_i)^2 (1/w_i) = \frac{S_w}{\Delta}$$

In summary,

$$\overline{A} = \frac{S_{wy} S_{wxx} - S_{wx} S_{wxy}}{S_w S_{wxx} - S_{wx}^2} \qquad \overline{\sigma}_A = \sqrt{\frac{S_{wxx}}{S_w S_{wxx} - S_{wx}^2}}$$

$$\overline{B} = \frac{S_w S_{wxy} - S_{wx} S_{wy}}{S_w S_{wxx} - S_{wx}^2} \qquad \overline{\sigma}_B = \sqrt{\frac{S_w}{S_w S_{wxx} - S_{wx}^2}}$$

4.3.3   Fitting $y = A + Bx$ in MATLAB

Example 4.2. Write a MATLAB function that fits a straight line $y = A + Bx$ through a dataset.

Solution 4.2. We can easily compute all the sums in MATLAB. Computing the sums by hand would be quite tedious.

```
function [ A sigA B sigB] = lineABx(xi,yi,sigi)
% [A sigA B sigB] = lineBx(xi,yi,sigi)
% Least sqaures fit to a line y = A + Bx

wi = sigi.^(-2);
Sw = sum(wi);
Swx = sum(wi.*xi);
Swxx = sum(wi.*xi.*xi);
Swy = sum(wi.*yi);
Swxy = sum(wi.*xi.*yi);
Delta = Sw*Swxx - Swx^2;

A = (Swy*Swxx - Swx*Swxy)/Delta;
B = (Sw*Swxy - Swx*Swy)/Delta;
sigA = sqrt(Swxx/Delta);
```

```
    sigB = sqrt(Sw/Delta);

end
```

Here is what the experimental flow looks like. You measure some points at each experimental setting. They have some spread and you measure that as the standard deviation. Then you have points and standard error bars. Then you go further and calculate the standard deviation of the mean and that is what you report and plot.

Example 4.3. Write a MATLAB code to display the reported values and their uncertainties (standard deviations of the mean).

Solution 4.3. `clear; figure(1);clf;`

```
% Generate some data
A = 10;
B = 3;
Ni = 15;
x = 1:2:15


for i = 1:length(x)
    ty = [];
    for j = 1:Ni
        ty(j) = A + B*x(i) + normrnd(0,3);
        plot(x(i),ty(j),'.r');
        hold on;
    end
    ey1(i) = std(ty);
    ey(i) = std(ty)/sqrt(Ni);
    y(i) = sum(ty)/Ni;
    % plot the distribution of the points
    yy=y(i)-12:0.01:y(i)+12;
    xx=x(i)+5/3/sqrt(2*pi).*exp(-(yy-y(i)).^2/(2*9));
    plot(xx,yy)
end
figure(2); clf;
errorbar(x,y,ey1,'.k','markersize',10);
figure(3);clf;
errorbar(x,y,ey,'.k', 'markersize',10);
xlabel('trial')
ylabel('measurement')
hold on;

[A sigA B sigB] = lineABx(x,y,ey)
```

```
y'
ey'
x=0:17;
y1 =    A + B.*x;
plot(x,y1, 'r');
axis( [ 0 17 0 1.1*max(y1)]);
xlabel('x','fontsize',20)
ylabel('mean \pm {\sigma}/N_i^{0.5}','fontsize',20);
figure(1);
axis( [ 0 17 0 1.1*max(y1)]);
xlabel('x','fontsize',20)
ylabel('measurement','fontsize',20);
figure(2);
axis( [ 0 17 0 1.1*max(y1)]);
xlabel('x','fontsize',20)
ylabel('mean \pm \sigma','fontsize',20);
```

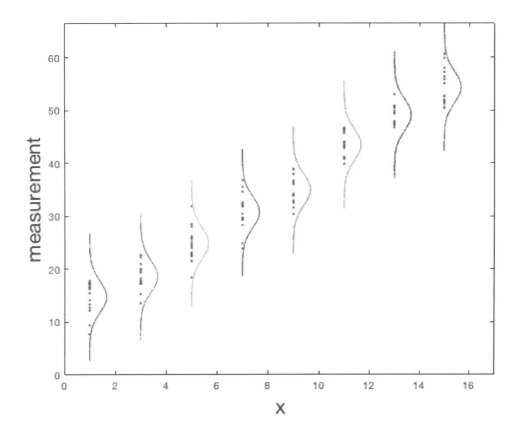

Figure 4.1: Raw data with 15 points at each setting

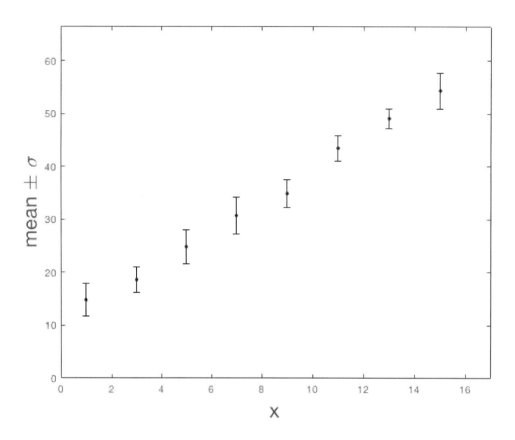

Figure 4.2: Calculation of standard deviation

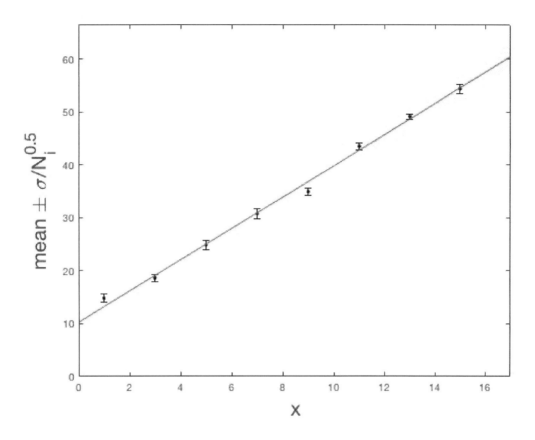

Figure 4.3: Calculation of standard deviation of the mean

4.3.4 Fitting $y = Bx$

For a model like $V = IR$, we want to fit a straight line forced through the origin, $y = Bx$. The calculus is similar.

$$\chi^2 = \sum_{i=1}^{N} \frac{(\overline{y}_i - Bx_i)^2}{\overline{\sigma}_i^2} = S_{wyy} - 2BS_{wxy} + B^2 S_{wxx}$$

Minimising for B using $\frac{d\chi^2}{dB} = 0$ we find

$$-2S_{wxy} + 2\overline{B}S_{wxx} = 0$$

$$\overline{B} = \frac{S_{wxy}}{S_{wxx}}$$

The propagation of error formula gives the uncertainty of B.

$$\overline{\sigma}_B^2 = \sum_{i=1}^{N} \left(\frac{\partial B}{\partial \overline{y}_i} \right)^2 \overline{\sigma}_i^2 = (1/S_{wxx}^2) \sum_{i=1}^{N} (w_i x_i)^2 w_i^{-1} = S_{wxx}/S_{wxx}^2$$

$$\overline{\sigma}_B = \frac{1}{\sqrt{S_{wxx}}}$$

Example 4.4. Write a MATLAB function to fit a straight line through the origin

Solution 4.4. We can evaluate the sums easily in MATLAB.

```
function [ B sigB ] = lineBx(xi,yi,sigi)
%[B sigB] = lineBx(xi,yi,sigi)
%Weightef least sqaured fit to a line y = Bx
wi = sigi.^(-2);
Swxy = sum(wi.*xi.*yi);
Swxx = sum(wi.*xi.*xi);
B = Swxy/Swxx;
sigB = 1/sqrt(Swxx);
end
```

Example 4.5. Fit some sample data that goes through the origin

Solution 4.5. clear; clf;
```
B = 3;
Ni = 5;
x = 1:2:15
for i = 1:length(x)
    tx = [];
    for j = 1:Ni
        tx(j) =  B*x(i) + normrnd(0,5);
    end
    ey(i) = std(tx)/sqrt(Ni);
    y(i) = sum(tx)/Ni;
end
figure(1);clf;
errorbar(x,y,ey,'.k', 'markersize',10);
xlabel('trial')
ylabel('measurement')
hold on;
[B sigB] = lineBx(x,y,ey)
[BU sigBU] = lineBxU(x,y)
y'
ey'
x=0:17;
y1 =    B.*x;
y2 = BU.*x;
plot(x,y1, 'g');
```

```
axis( [ 0 17 0 1.1*max(y1)]);
hold on
plot(x,y2, 'r')
```

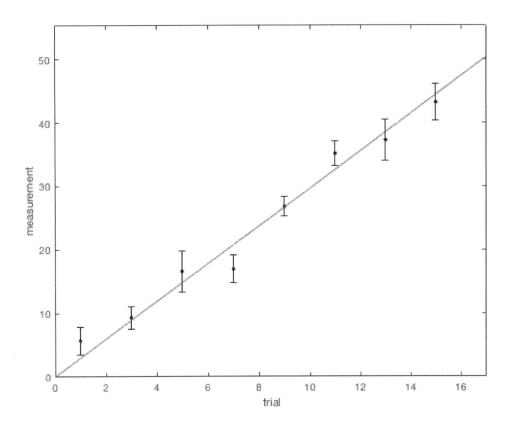

Figure 4.4: Curve through the origin, $y = Bx$

4.4   Fitting $y = A$

4.4.1   The weighted mean

When the data follow $y = A$ we expect them to line up on a horizontal line or a constant function. We are taking a weighted average of the $\overline{y}_i$. A is called the weighted mean, since the points are weighted by $1/\sigma_i^2$ in the chi squared formula given by

$$\chi^2 = \sum_{i=1}^{N} \frac{(\overline{y}_i - A)^2}{\sigma_i^2}$$

$$\chi^2 = \sum_{i=1}^{N} \frac{\overline{y}_i^2 - 2A\overline{y}_i + A^2}{\sigma_i^2}$$

$$\chi^2 = S_{wyy} - 2AS_{wy} + A^2 S_w$$

Now we minimise according to $\dfrac{d}{dA}\chi^2 = 0$.

$$0 = -2S_{wy} + 2\overline{A}S_w$$

$$\overline{A} = \frac{S_{wy}}{S_w}$$

$$\overline{A} = \overline{y}_w = \frac{w_1\overline{y}_1 + w_2\overline{y}_2 + \ldots + w_N\overline{y}_N}{w_1 + w_2 + \ldots + w_N}$$

We use the symbol $\overline{y}_w = \overline{A}$ to represent the weighted mean.

4.4.2   Uncertainty in the weighted mean

We can find the uncertainty in the weighted mean by using the propagation of error formula.

$$\overline{\sigma}_w^2 = \sum_{i=1}^{N}\left(\frac{\partial B}{\partial \overline{y}_w}\right)^2 \overline{\sigma}_i^2 = (1/S_w^2)\sum_{i=1}^{N} w_i^2 \frac{1}{w_i} = 1/S_w$$

$$\overline{\sigma}_w = \frac{1}{\sqrt{S_w}} = \frac{1}{\sqrt{\displaystyle\sum_{i=1}^{N}\frac{1}{\overline{\sigma}_i^2}}}$$

4.4.3   Combining a series of results

The weighted mean provides a way of combining different experimental results of the same quantity. For example, several groups measure the quantity G. Then you combine the results using the weighted mean to determine the accepted value. The experiments with smaller uncertainties are going to be closer to the accepted value, because of how the sum is weighted. This is how the fundamental constants and uncertainties are calculated from all the available experiments.

4.4.4   Discrepancy

Discrepancy is how far apart the results are for two separate measurements of the same thing. One usually expects the range of the uncertainties to overlap. Otherwise one would expect that there is a problem with one or more of the experiments. When several different experiments give results with discrepancy there is definitely something to figure out or more experiments are warranted.

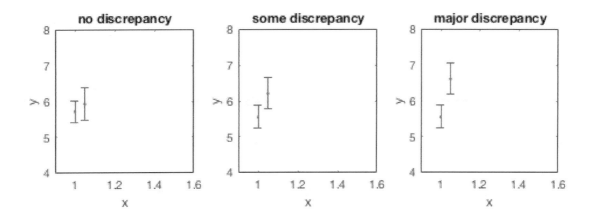

Figure 4.5: Illustrating discrepancy

Example 4.6. Show on a diagram the varying degrees of discrepancy.

Solution 4.6. Making up some random data we can show how the they overlap.

```
x1 = [1 1.05]
y1 = [ 5.72 5.93]
sdm1 = [0.3 0.44]

x2 = [1 1.05]
y2 = [ 5.55 6.21]
sdm2 = [ 0.32 0.44]

x3 = [1 1.05]
y3 = [ 5.55 6.61]
sdm3 = [ 0.32 0.44]

figure(1);clf;
subplot(1,3,1)
errorbar(x1,y1,sdm1,'.');
axis([ 0.9 1.6 4 8]);
axis square
hold on;
xlabel('x');
ylabel('y');
title('no discrepancy')
subplot(1,3,2);
errorbar(x2,y2,sdm2,'.');
axis([ 0.9 1.6 4 8]);
axis square
```

```
xlabel('x');
ylabel('y');
title('some discrepancy')
subplot(1,3,3);
errorbar(x3,y3,sdm3,'.');
axis([ 0.9 1.6 4 8]);
axis square
xlabel('x');
ylabel('y');
title('major discrepancy')
```

4.4.5   Weighted mean in MATLAB

Example 4.7. Write a MATLAB function to calculated the weighted mean and its uncertainty.

Solution 4.7. Implementing the theory we have

```
function [ A sigA ] = lineA( xi, yi, sigi )
% [A sigA] = lineA(xi, yi, sigi)
% Least sqaures fit to a constant y = A

wi = sigi.^(-2);
Sw = sum(wi);
Swy = sum(wi.*yi);
A = Swy/Sw;
sigA = 1/sqrt(Sw);

end
```

Example 4.8. Simulate some data illustrating the weighted mean.

Solution 4.8. We will assume the weighted

```
x = 1:10
for i = 1:10
    ey(i) = 0.3 + abs(normrnd(0,2));
    y(i) = normrnd(10, ey(i));
end
figure(1);clf;
errorbar(x,y,ey,'.k', 'markersize',10);
xlabel('trial')
ylabel('measurement')
hold on;
```

```
[A sigA] = lineA( x,y,ey)
y'
ey'

x1 = [ 0 11];
y1 = [ A  A];
y2 = [ (A +sigA) (A + sigA) ];
y3 = [ (A -sigA) (A- sigA) ];
plot(x1,y1, 'r');
axis( [ 0 11 0 15])
plot(x1,y2, '-.r');
plot(x1,y3, '-.r');

>> A = 9.855394364614574
>> sigA = 0.324748493970210
```

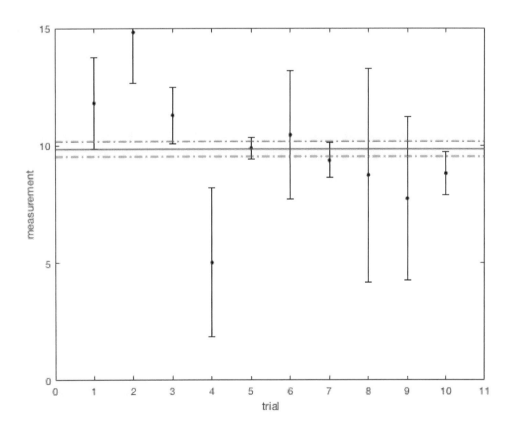

Figure 4.6: Weighted mean example y = A

4.5   Unweighted fits and outlying points

4.5.1   Unweighted linear fits

You should always try to repeat each measurement before you model the data as a linear fit. This is so you can determine the uncertainties and properly weight the data. There may be situations however where you cannot repeat the measurements. You may not have sufficient time or the measurements are too expensive. We can still do least squares curve fitting in this situation if we assume all the measurements have more or less the same uncertainty and deviation away from a straight line.

$$\chi^2 = \sum_{i=1}^{N} \frac{(\overline{y}_i - A - Bx_i)^2}{\sigma^2}$$

To determine σ which is the standard deviation because you only have one measurement, we assume the numerator is of order σ^2 for each point then

$$N \approx \sum_{i=1}^{N} \frac{(\overline{y}_i - \overline{A} - \overline{B}x_i)^2}{\sigma^2}$$

and

$$\sigma = \sqrt{\sum_{i=1}^{N} \frac{(\overline{y}_i - \overline{A} - \overline{B}x_i)^2}{N}}$$

After we find $\overline{A}$ and $\overline{B}$, we can find σ in order to find the uncertainties in the parameters $\overline{A}$ and $\overline{B}$. We can use the previous analysis where the weights are just $w_i = \dfrac{1}{\sigma^2}$. There will then be some cancelation depending on the model or parameters.

Example 4.9. Write a MATLAB function to do an unweighted least squares fit to the line $y = A + Bx$.

Solution 4.9. Implementing the formulas we have

```
function [ A sigA B sigB] = lineABxU( xi, yi )
%[A sigA B sigB] = lineBx(xi, yi, sigi)
%    Least sqaures fit to a line y = A + Bx
S = length(xi);
Sx = sum(xi);
Sxx = sum(xi.*xi);
Sy = sum(yi);
Sxy = sum(xi.*yi);
Delta = S*Sxx - Sx^2;
A = (Sy*Sxx - Sx*Sxy)/Delta;
B = (S*Sxy - Sx*Sy)/Delta;
sigma = sqrt(1/length(xi)*sum((yi - A - B.*xi).^2));
```

```
    sigA = sigma*sqrt(Sxx/Delta);
    sigB = sigma*sqrt(S/Delta);
    end
```

4.5.2 Outliers and linear curve fits

Example. Show the effect of an outlying point when a linear fit is weighted or unweighted. We will use simulated data like before but where the last point is an outlier. We will show the effect in an unweighted fit and a weighted fit which is more robust and fits the data better.

```
clear;
format long;
x = 1:10
y = [ 13.69 14.9 17.1 24.4 25.4 26.6 29.48 32.0 37.36 48.0 ]
sigy = [ 0.58 1.0 1.0 1.2 1.6 1.4 0.86 1.8 0.84 8.1 ]

figure(1); clf;
errorbar(x,y,sigy,'.k');
hold on;
axis( [ 0 11 0 1.3*max(y)]);
xlabel('x','fontsize',20);
ylabel('y','fontsize',20);
title('Effect of an outlying point on unweighted fit','fontsize',18);
[A1 sigA1 B1 sigB1] = lineABx(x,y, sigy);
[A2 sigA2 B2 sigB2 ] = lineABxU(x, y);

x1 = [0 11]
y1 = [ A1 (A1 +B1*11)];
plot(x1,y1,'k-');
x2 = [0 11]
y2 = [ A2 (A2 +B2*11)];
plot(x2,y2,'k-.');
```

Effect of an outlying point on unweighted fit

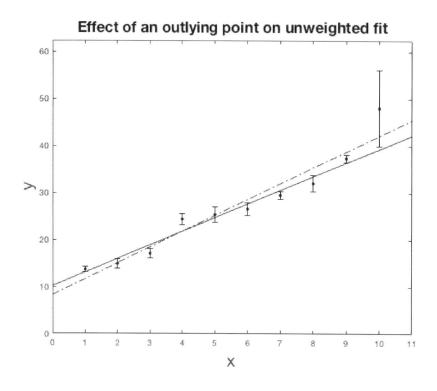

Figure 4.7: Effect of Outliers

Model $y = 10 + 3x + normrnd(0,3)$, with five samples, standard deviation of the mean used and an outlying point. The outlying point shifts the slope of the unweighted fit, but does not effect the weighted fit too much. The outlying point has a larger error bar typically indicating that something is wrong or statistically unlikely to occur.

4.5.3   Rejection of data, Chauvenet's criterion

If you want to clean up your graphs you can nuke any points that are more than 3 standard deviations away from the mean. This is called Chauvenet's criterion. Some scientists disagree that you shouldn't throw out any data because it might signify something real. The most objective way is to repeat the measurement at that point many many times. Repetition should bring it back in line towards the reasonable value.

Sometimes we do make mistakes like recording a number incorrectly in our lab notebook. You should find that you are not rejecting data that often especially when you deal with datasets with a lot of variation. You should establish what the criterion for rejecting data is before you make the measurements in any event, then you cannot be accused of misrepresentation.

4.6   Linearization of nonlinear models

4.6.1   Linear and nonlinear models

In least squares fitting the models fall into two categories, linear and nonlinear. The linear models have analytical solutions. The nonlinear models must be solved numerically or transformed into a linear model through propagation of error. The most general linear model with M parameters can be written as

$$f(x_i; \mathbf{a}) = a_1 X_1(x_i) + a_2 X_2(x_i) + \ldots + a_M X_M(x_i)$$

A useful linear model is a polynomial of degree $M - 1$, this includes constant functions, straight lines, and parabolas. You can fit most curves with a polynomial if you choose a large number for M. This is because most formulas have a Taylor series that you can take advantage of.

$$f(x_i; \mathbf{a}) = a_1 + a_2 x_i + \ldots + a_M x_i^{M-1}$$

You can have other models that are linear, but they appear a little goofy. For example, you would need to know something about the frequencies a priori if you wanted to fit a Fourier series. The following formula is also linear

$$f(x_i; \mathbf{a}) = a_1 \sin(x_i) + a_2 \sin(2x_i) + \ldots + a_M \sin(Mx_i)$$

Sometimes data following a nonlinear relationship may be transformed to a linear model. One can then solve for the transformed parameters exactly and convert them to the originals back by propagation of error.

4.6.2   Lineweaver-Burke plot

Here is a well known example from biochemistry for the velocity of a chemical reaction illustrating linearization.

$$V = \frac{V_{max}[S]}{K_m + [S]}$$

This equation can be rewritten as

$$\frac{1}{V} = \frac{K_m}{V_{max}} \frac{1}{[S]} + \frac{1}{V_{max}}$$

We start with $([S]_i, V_i, \sigma_{V_i})$ as the set of information. σ_{V_i} is the standard deviation of the mean of the repeated measurement of V_i. The transformed variables are

$$x_i = 1/[S]_i \quad y_i = 1/V_i \quad \sigma_i = \frac{\sigma_{V_i}}{V_i^2}$$

$$a_2 = \frac{K_m}{V_{max}} \qquad a_1 = \frac{1}{V_{max}}$$

Propagation of error for a single variable has been used to find σ_i. One can then weighted fit to a line to find a_1, a_2, σ_{a_1}, and σ_{a_2}. K_m and V_{max} can then be found also by propagation of error. This transformation generates a linear plot called the Lineweaver-Burke plot.

4.6.3   Linearization of exponential data

Another straightforward example of linearization involves exponential decay.

$$N(t) = N_0 e^{-kt}$$

One has the set of information (t_i, N_i, σ_{N_i}) where σ_{N_i} are the standard deviations of the mean. Now take the logarithm of both sides.

$$\log(N_i) = -kt_i + \log(N_0)$$

Now we transform

$$x_i = t_i \quad y_i = \log(N_i) \quad \sigma_i = \sigma_{N_i}/N_i$$
$$a_1 = \log(N_0) \quad a_2 = -k$$

If the error bars were originally similar for most points on the linear plot they will be stretched out at longer times after transformation. After finding a_1 and σ_{a_1}, you can use error propagation to find N_0 and its uncertainty.

$$\sigma_{N_0} = e^{a_1}\sigma_{a_1}$$

Example 4.10. Fit some simulated exponentially decaying data with linear least squares by linearizing the model.

Solution 4.10. We will make some data that decays from 100 with a rate of 1. If you run the code you can see the curve fit works. On a sample run I found, 100.01(12) and 0.9973(48) for the fit parameters and uncertainties. We use the method of linearization to convert exponential data into a linear model that can be fit with the previous subroutines.

```
clear;
figure(1);clf;
xi = 0.0:0.1:3

N0 = 100;
lambda = 1
for i = 1:length(xi)
    yi(i) = N0*exp(-lambda*xi(i));
    ei(i) = normrnd(0,3);
    yi(i) = abs(yi(i) + ei(i));
end

errorbar(xi,yi,ei);
xlabel('x');
ylabel('N');
axis([0 3 0 125]);

figure(2);clf;
```

```
zi = log(yi);
si = ei./yi;
errorbar(xi,zi,si,'.');

[A sigA B sigB ] = lineABx(xi, zi, si);
A
sigA
B
sigB
XX = 0:0.1:3
YY = A + XX.*B;
hold on
plot(XX,YY);

myN = exp(A)
mysN = exp(A)*sigA
% ln N = ln N_0 + -Lx
myL = B
mysL = sigB
axis([0 3 0 5]);
```

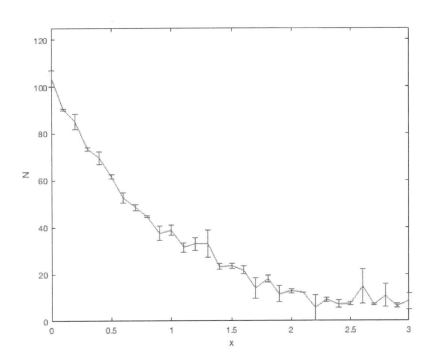

Figure 4.8: Exponential Decay

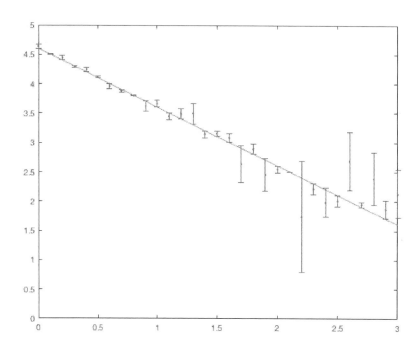

Figure 4.9: Linear Transformation

4.7   Generalized least squares

4.7.1   Generalized least squares models

Now we would like to solve the most general linear model analytically to find the best fit parameters. The model is.

$$f(x_i; a_1, a_2, ..., a_M) = f(x_i; \mathbf{a}) = \sum_{k=1}^{M} a_k X_k(x_i)$$

$$\chi^2 = \sum_{i=1}^{N} \frac{(\overline{y}_i - \sum_{k=1}^{M} a_k X_k(x_i))^2}{\overline{\sigma}_i^2}$$

To minimise χ^2, we find k equations according to $\dfrac{\partial}{\partial a_k}\chi^2 = 0$. These "normal equations" give the optimal fit parameters $\overline{\mathbf{a}}$.

$$\sum_{i=1}^{N} \left[\left(\overline{y}_i - \sum_{j=1}^{M} \overline{a}_j X_j(x_i) \right) \frac{X_k(x_i)}{\overline{\sigma}_i^2} \right] = 0$$

Taking terms on each side of the equal sign.

$$\sum_{i=1}^{N} \frac{\overline{y}_i X_k(x_i)}{\overline{\sigma}_i^2} = \sum_{i=1}^{N} \sum_{j=1}^{M} \overline{a}_j X_j(x_i) \frac{X_k(x_i)}{\overline{\sigma}_i^2}$$

It is useful to do some linear algebra to solve these equations for the parameters and their uncertainties. The "design matrix," A_{ij}, is defined as the N row and M column matrix

$$A_{ij} = \frac{X_j(x_i)}{\overline{\sigma}_i}$$

Second we define the column vector $\mathbf{B}$ of M rows.

$$B_k = \sum_{i=1}^{N} \frac{\overline{y}_i X_k(x_i)}{\overline{\sigma}_i^2}$$

A third $M \times M$ matrix, D_{jk} has inverse C_{jk}.

$$D_{jk} = \sum_{i=1}^{N} \frac{X_j(x_i)X_k(x_i)}{\overline{\sigma}_i^2} \rightarrow \mathbf{D} = \mathbf{A}^T\mathbf{A}$$

$$\mathbf{C} = \mathbf{D}^{-1}$$

4.7.2 Generalized least squares fit parameters

Now the normal equations can be rewritten as

$$\mathbf{D}\overline{\mathbf{a}} = \mathbf{B}$$

To solve for the fit parameters $\overline{\mathbf{a}}$ then numerically compute the inverse

$$\overline{\mathbf{a}} = \mathbf{D}^{-1}\mathbf{B} = \mathbf{C}\mathbf{B}$$

$$\overline{a}_j = \sum_{k=1}^{M} C_{jk} \left[\sum_{i=1}^{N} \frac{y_i X_k(x_i)}{\overline{\sigma}_i^2} \right]$$

4.7.3 Generalized lease squares parmaeter uncertainties

Now, we can find the uncertainty in the parameters by the propagation of error formula.

$$\overline{\sigma}_{a_j}^2 = \sum_{i=1}^{N} \left(\frac{\partial a_j}{\partial \overline{y}_i} \right)^2 \overline{\sigma}_i^2$$

$$\frac{\partial a_j}{\partial \overline{y}_i} = \sum_{k=1}^{M} \frac{C_{jk} X_k(x_i)}{\overline{\sigma}_i^2}$$

$$\overline{\sigma}_{a_j}^2 = \sum_{k=1}^{M} \sum_{l=1}^{M} C_{jk} C_{jl} \left(\sum_{i=1}^{N} \frac{X_k(x_i)X_l(x_i)}{\overline{\sigma}_i^2} \right)$$

$$\overline{\sigma}_{a_j}^2 = \sum_{k=1}^{M} \sum_{l=1}^{M} C_{jk} C_{jl} D_{lk}$$

But $\displaystyle\sum_{l=1}^{M} C_{jl} D_{lk} = \delta_{jk}$ the Kroenecker delta because $\mathbf{C}$ is the inverse of $\mathbf{D}$.

$$\delta_{ij} = \begin{cases} 1, & \text{if } i = j, \\ 0, & \text{if } i \neq j. \end{cases}$$

so the sum reduces to

$$\overline{\sigma}^2_{a_j} = C_{jj}$$

These are the uncertainties in the fitting parameters. The off diagonal terms of $\mathbf{C}$ also give the covariances. If two different groups use the least squares theory on a linear model they should arrive at the same results for the fitting parameters and their uncertainties which are derived here. To summarise the results for linear least squares models

$$\overline{a}_j = \sum_{k=1}^{M} C_{jk} \left[\sum_{i=1}^{N} \frac{\overline{y}_i X_k(x_i)}{\overline{\sigma}_i^2} \right] \qquad \overline{\sigma}_{a_j} = \sqrt{C_{jj}}$$

4.8   Least squares polynomial

4.8.1   Why fit a polynomial

Sometimes if you don't have a good model, it is best to fit a polynomial of sufficient degree through your data. This has several advantages. First it interpolates the data. You can take the interpolating polynomial and exactly integrate or differentiate it. So it will allow you to interact with your data quantitatively over the entire range you have measured. Also, most functions can be approximated as a polynomial. So, in certain instances you are just seeing the Taylor expansion of a more general function.

4.8.2   Least squares polynomial theory

We have for

$$D_{jk} = \sum_{i=1}^{N} x_i^{j-1} x_i^{k-1}$$

$$B_k = \sum_{i-1}^{N} x_i^{k-1} y_i \sigma_i$$

$$\overline{\mathbf{a}} = \mathbf{D}^{-1} \mathbf{B}$$

$$\mathbf{C} = \mathbf{D}^{-1}$$

$$\overline{\sigma}_{a_i} = \sqrt{C_{kk}}$$

4.8.3   Implementing polynomial linear least squares in MATLAB

Example 4.11. Implement a polynomial least squares fitting function in MATLAB.

Solution 4.11. We can tweak our generalized least squares function a little bit.

```
function [ A sigA ] = genpolyfit( M, xi, yi, sigi )
%   [A sigA ] = genpolyfi( M, xi, yi, sigi)
%   Least sqaures polynomial fit to degree M-1
%

for k = 1:M
    X(k,:) = xi.^(k-1);
end

for j = 1:M
    for k = 1:M
        D(j,k) = sum(sigi.^(-2).*X(j,:).*X(k,:));
    end
end

for k = 1:M
    B(k) = sum(X(k,:).*yi.*sigi.^(-2));
end

A = inv(D)*B';
C = inv(D);
for k = 1:M
    sigA(k) = sqrt(C(k,k));
end
```

Example 4.12. Simulate some data that can be fit by a parabola.

Solution 4.12. We can use the least square polynomial subroutine.

```
clear; clf;
A = 2;
B = 5;
C = 3;
Ni = 10;
xi = -10:10
for i = 1:21
    yj = [];
    for j = 1:Ni
        yj(j) = A + B*xi(i) + C*xi(i).^2 + normrnd(0,10);
    end
```

```
    yi(i) = sum(yj)/length(yj);
    si(i) = std(yj,1)/sqrt(Ni);
end
figure(1);clf;
errorbar(xi,yi,si,'.k', 'markersize',10);
xlabel('trial')
ylabel('measurement')
hold on;
[A sigA] = genpolyfit(3,xi,yi,si);

x1 = -11:0.1:11;
y1 = A(3).*x1.^2 + A(2).*x1 + A(1);
plot(x1,y1, 'r');
axis( [ -11 11 -50 1.1*max(y1)]);
A
sigA'
```

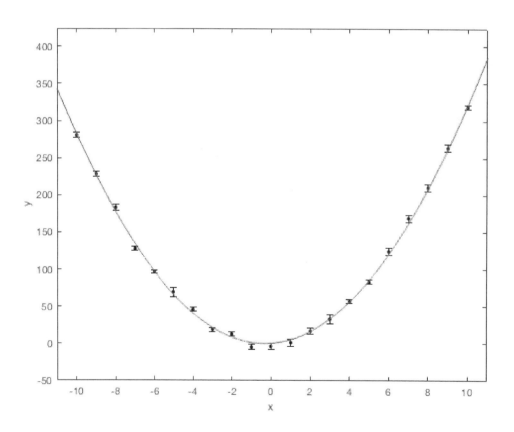

Figure 4.10: Curve fit to a parabola

4.9 Summary

- Model with datasets $\{x_i, \overline{y}_i, \overline{\sigma}_i\}$

- Linear model $f(x_i; \mathbf{a}) = \sum\limits_{j=1}^{M} a_j X_j(x_i)$

- $\chi^2 = \sum\limits_{i=1}^{N} \dfrac{(\overline{y}_i - f(x_i; \mathbf{a}))^2}{\overline{\sigma}_i^2}$

- $y = A + Bx$

$$\overline{A} = \frac{S_{wy}S_{wxx} - S_{wx}S_{wxy}}{S_w S_{wxx} - S_{wx}^2} \qquad \overline{\sigma}_A = \sqrt{\frac{S_{wxx}}{S_w S_{wxx} - S_{wx}^2}}$$

$$\overline{B} = \frac{S_w S_{wxy} - S_{wx}S_{wy}}{S_w S_{wxx} - S_{wx}^2} \qquad \overline{\sigma}_B = \sqrt{\frac{S_w}{S_w S_{wxx} - S_{wx}^2}}$$

- $y = Bx$

$$\overline{B} = \frac{S_{wxy}}{S_{wxx}} \qquad \overline{\sigma}_B = \frac{1}{\sqrt{S_{wxx}}}$$

- $y = A$, Weighted mean

$$\overline{A} = \frac{w_1 \overline{y}_1 + w_2 \overline{y}_2 + \ldots + w_N \overline{y}_N}{w_1 + w_2 + \ldots + w_N} \qquad \overline{\sigma}_w = \frac{1}{\sqrt{S_w}} = \frac{1}{\sqrt{\sum\limits_{i=1}^{N} \dfrac{1}{\overline{\sigma}_i^2}}}$$

- Unweighted curve fits

$$w_i = \frac{1}{\sigma^2} \qquad \sigma = \sqrt{\sum\limits_{i=1}^{N} \frac{(\overline{y}_i - f(x_i; \overline{\mathbf{a}}))^2}{N}}$$

- Generalized least squares fit parameters $\overline{\mathbf{a}} = \mathbf{D}^{-1}\mathbf{B} = \mathbf{CB}$

- Generalized least squares uncertainties $\overline{\sigma}_{a_j} = \sqrt{C_{jj}}$

Chapter 5

Nonlinear least squares

5.1 The minimisation of nonlinear data

We have our function χ^2 which we minimise in least squares theory.

$$\chi^2 = \sum_{i=1}^{N} \frac{(\overline{y}_i - f(x_i; \mathbf{a}))^2}{\overline{\sigma}_i^2}$$

We showed that when f is a linear function according to

$$f(x_i; \mathbf{a}) = a_1 X_1(x_i) + a_2 X_2(x_i) + \cdots + a_M X_M(x_i)$$

Then χ^2 has a unique global minimum and we can analytically derive expressions for $\mathbf{a}$ and $\overline{\sigma}_{a_i}$. When f is not of the above form then it is nonlinear. When you take the derivative you get a nonlinear equation which may have no analytical solution There is no general solution for the set of $\mathbf{a}$ that minimise χ^2 in the nonlinear case. One has to resort to different methods. One typically resorts to iterative methods that requires an initial guess for the set $\mathbf{a}$. When one applies the procedure on the initial guess the subsequent value is closer to the minimum call it $\mathbf{a}(1)$. One repeats this many times and hopefully the final coordinate $\mathbf{a}(n)$ is very near the global minimum.

Throughout this chapter we will develop different curve fitting methods. We can choose a nonlinear model to try to test the methods on. A common model is an oscillating mass on a spring with damping.

$$m\ddot{y} + b\dot{y} + km = 0$$

For a certain choice of constants, the motion is called critically damped and obeys the following equation

$$y = (A + Bt)\exp(-Ct)$$

So we are going to make some simulated data of that function where we measure the curve five times and average. Then we will try the different iterative methods in this chapter to find the best fit parameters.

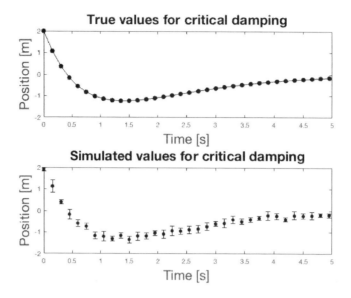

Figure 5.1: Exponential Decay

```
clear;figure(1);clf;
% Model data
A = 2;
B = -5;
C = 1;

xi = 0:0.15:5
ymodel = (A+B.*xi).*exp(-C.*xi);

subplot(2,1,1)
plot(xi,ymodel,'.-k','markersize',20)
hold on;
xlabel( 'Time [s]','fontsize',20);
ylabel( 'Position [m]','fontsize',20);
title('True values for critical damping','fontsize',20);

for i = 1:length(xi)
    ty = []
    for j = 1:5
        ty(j) = (A + B.*xi(i)).*exp(-C.*xi(i)) + normrnd(0,0.17)
    end
    baryi(i) = mean(ty);
    barsigi(i) = std(ty,1);
end

subplot(2,1,2)
errorbar(xi,baryi, barsigi,'.k','markersize',15);
```

```
xlabel( 'Time [s]','fontsize',20);
ylabel( 'Position [m]','fontsize',20);
title('Simulated values for critical damping','fontsize',20);

save modeldata xi baryi barsigi
```

5.2 Least squares uncertainties

If you can find the optimal parameters, then it is straightforward to determine their uncertaintites. Remember from D_{ij} from the generalised least squares theory. We used it to find $\mathbf{C} = \mathbf{D}^{-1}$.

$$D_{jk} = \sum_{i=1}^{N} \frac{X_j(x_i)X_k(x_i)}{\overline{\sigma}_i^2} \rightarrow \mathbf{D} = \mathbf{A}^T\mathbf{A}$$
$$\mathbf{C} = \mathbf{D}^{-1}$$

The parameter uncertainties were given by

$$\sqrt{C_{jj}} = \overline{\sigma}_{a_j}$$

There is another clever way to calculate D_{jk}

$$D_{jk} = (1/2)\frac{\partial^2 \chi^2}{\partial a_j \partial a_k}$$

We see this is one half the Hessian matrix.

We can use the same method for nonlinear models.

$$D_{jk} = \sum_{i=1}^{N} \frac{1}{\overline{\sigma}_i^2}\left[\left(\frac{\partial f}{\partial a_j}\right)\left(\frac{\partial f}{\partial a_k}\right) + (f - \overline{y}_i)\frac{\partial^2 f}{\partial a_j \partial a_k}\right]$$

and then you can work out the parameter uncertainties. We can demonstrate this method works for the linear models where we already have some exact results.

Example 5.1. The formula for χ^2 when we fit the line through the origin was

$$\chi^2 = S_{wyy} - 2BS_{wxy} + B^2 S_{wxx}$$

Find the uncertainty of $\overline{B}$.

Solution 5.1. We can easily find $\mathbf{D}$ it is a 1 by 1 matrix.

$$\mathbf{D} = (1/2)\frac{d^2}{dB^2}\chi^2 = S_{wxx}$$

Since $\mathbf{C}$ is the inverse of $\mathbf{D}$ we have

$$\mathbf{C} = \frac{1}{S_{wxx}}$$

Therefore

$$\overline{\sigma}_B = \frac{1}{\sqrt{S_{wxx}}}$$

This is exactly what we got from before using the propagation of error formula.

Example 5.2. The χ^2 for fitting the line $y = A + Bx$ is given by

$$\chi^2 = A^2 S_w + 2AB S_{wx} + B^2 S_{wxx} - 2A S_{wy} - 2B S_{wxy} + S_{wyy}$$

Find **D** and its inverse **C**. Show that

$$\overline{\sigma}_A = \sqrt{C_{AA}} \qquad \overline{\sigma}_B = \sqrt{C_{BB}}$$

give the same results as the propagation of error formula used in the previous chapter.

Solution 5.2. We can just calculate the partial derivatives.

$$\mathbf{D} = (1/2)\frac{d^2}{dB^2}\chi^2 = S_{wxx}$$

$$D_{11} = S_w \quad D_22 = S_{wxx} \quad D_{12} = S_{wx} \quad D_{21} = S_{wx}$$

The matrix **D** can be written as

$$D_{jk} = \begin{pmatrix} S_w & S_{wx} \\ S_{wx} & S_{wxx} \end{pmatrix}$$

The inverse of **D** is **C**. $\Delta = det(\mathbf{D})$.

$$C_{11} = S_{wxx}/\Delta \qquad C_{22} = S_w/\Delta$$

We get the same uncertainties like before when we used propagation of error.

Example 5.3. χ^2 for the weighted mean is given by

$$\chi^2 = S_{wyy} - 2A S_{wy} + A^2 S_w$$

Find **D** then **C** and show that $\overline{\sigma}_w = 1/\sqrt{S_w}$

Solution 5.3. This is one dimensional so it is easy $D_{11} = S_w$. The **C** matrix is just

$$\mathbf{C} = (1/S_w)$$

So the uncertainty is $\sqrt{C_{11}} = 1/\sqrt{S_w}$. Piece of cake!

5.3 Newton's method

We will explore Newton's method because it is a simple example of an iterative method. We generally use iterative methods to minimise χ^2 when f is nonlinear. Newton's method is an iterative method to find the root of a function.

Consider the following diagram figure x of $f(x)$. The quantities can be related by the following equation.

$$f'(x_1) = f(x_1)/(x_1 - x_2)$$
$$x_2 = x_1 - \frac{f(x_1)}{f'(x_1)}$$

So if we know x_1 the initial guess, we can find x_2. Then we can plug back in x_2 in the next iteration to find x_3 and so on. We get closer and closer to the point where $f(x) = 0$ with each iteration

$$x_{n+1} = x_n - \frac{f(x_n)}{f'(x_n)}$$

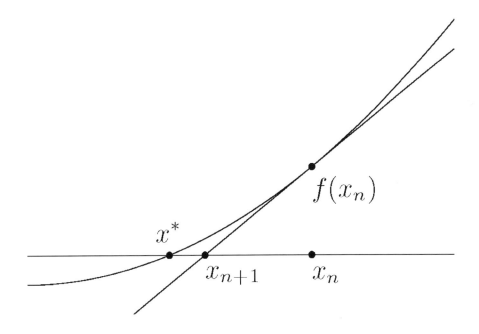

Figure 5.2: Newton's method

Example 5.4. Find the square root of two to high precision.

Solution 5.4. We can try out Newton's method to calculate the square root of 2. If we have the equation,

$$x^2 = 2$$

Then we can write $f(x) = x^2 - 2$. This function goes to zero at $\sqrt{2}$ so we can apply Newton's method.

$$x_{n+1} = x_n - \frac{x_n^2 - 2}{2x_n}$$

```
clear; format long;
N = 10;
x(1) = 2.
for i = 1:N
    x(i+1) = x(i) - (x(i)^2 -2)/(2*x(i));
end
```

If we look at the contents of x we find,

n	x_n	
1	2	
2	1.500	
3	1.416666666666667	
4	1.414215686274510	The update function just produces fractions or rational numbers
5	1.414213562374690	
6	1.414213562373095	
7	1.414213562373095	

if you choose an initial guess like 2. In old times, before the calculator one could calculate the square root of any number to sufficient accuracy by finding the fraction, then doing long division in the last step. It is quite a good method. 15 digits are perfect after the 6th iteration.

There are a number of things that can go wrong with Newton's method. Each successive update may result in oscillation or the updates may be directed to another solution you are not looking for. The best way to find a root is to make a plot of $f(x)$ near the root and plan your initial guess such that it climbs down the hill to the root avoiding any bumps or being too far away from the minimum. Also your function may have a restricted domain or singularities where division by zero occurs. These will obviously crash Newton's method.

5.4   Brute force grid method

If you have the computational power, then there is a brute force approach to curve fitting. This method is called the grid method. Suppose from the graph of some linear data you can tell that A

is between 3 and 4 and B is between 0.5 and 1.5. Then you can calculate χ^2 for every value in between to 0.01 or 0.001 resolution. You then just find the least one by direct search in MATLAB. Of course those might take 10,000 or 100,000 calculations but you will get the same answer as the analytical result. You can do a coarse grid to find local minima and then do a fine grid to find the global minimum. One advantage of the grid method is that you don't need to know much theory and it is pretty obvious how it works. If you cannot narrow the parameter space to include the global minimum, then you might require too many calculations to perform the optimisation.

We will now study some iterative methods that can be used to minimise chi squared. Grid method is not really an iterative method like the rest. You try to cast a net over the parameter surface. You see which loop of the net produces the minimum, then you cast a finer net over that loop and rapidly converge towards a minimum. This method is good because you get out of it what you expect. When you have lots of parameters it might not be possible to cast a fine enough net to progress because there are too many combinations.

5.5 Important matrix math

We can represent χ^2 in terms of vectors. Let

$$\chi^2 = \sum_{i=1}^{N} \frac{(\bar{y}_i - f(x_i; \mathbf{a}))^2}{\bar{\sigma}_i^2}$$

We define the residual vector.

$$\mathbf{r} = \bar{y}_i - f_i$$

We define the weight matrix which is diagonal.

$$W_{ij} = \frac{1}{\bar{\sigma}_i^2} \delta_{ij}$$

So we can write χ^2 as

$$\chi^2(\mathbf{a}) = \mathbf{r}^{\mathbf{T}} \mathbf{W} \mathbf{r}$$

We go back and forth between column and row vectors using the transpose property. Some properties of transpose are

$$(\mathbf{A}^{\mathbf{T}})^{\mathbf{T}} = \mathbf{A}$$
$$\mathbf{A}^{\mathbf{T}} = \mathbf{A} \rightarrow \mathbf{A} \quad symmetric$$
$$(\mathbf{A} + \mathbf{B})^T = \mathbf{A}^T + \mathbf{B}^T$$
$$(\mathbf{AB})^{\mathbf{T}} = \mathbf{B}^{\mathbf{T}} \mathbf{A}^{\mathbf{T}}$$

When you multiply an NxM matrix with an MxL matrix you get an NxL matrix. You can use size(A) command in MATLAB to see how your matrices are arranged. You can use the transpose operator A' to convert a column vector to a row vector and vice versa. You may have to check your expression if you haven't kept track of your transposes.

We will be taking partial derivatives of χ^2 with respect to the parameters which is the Jacobian matrix.

$$\frac{\partial \mathbf{f}}{\partial \mathbf{a}} = \mathbf{J}$$

$$\frac{\partial \mathbf{r}}{\partial \mathbf{a}} = -\mathbf{J}$$

We also have the Hessian matrix which is defined as

$$H_{ij} = \frac{\partial^2 \chi^2}{\partial a_j \partial a_k}$$

Here is an important result from Matrix calculus for a symmetric matrix $\mathbf{W}$.

$$\frac{\partial(\mathbf{h}^\mathbf{T}\mathbf{W}\mathbf{h})}{\partial \mathbf{x}} = 2\mathbf{h}^\mathbf{T}\mathbf{W}\frac{\partial \mathbf{h}}{\partial \mathbf{x}}$$

5.6   Gradient decent

Imagine χ^2 as a multidimensional hyper-surface of the possible fit parameters, $\mathbf{a} = (a_1, a_2, ..., a_N)$. If you choose a set of $\mathbf{a}$. We can use the result from vector calculus that $-\nabla\chi^2$ will point towards a local minimum. We can then step in that direction and repeat the process. We take an arbitrary step size of $\gamma = 0.00001$. We take the gradient of chi squared.

$$\nabla\chi^2 = -2\mathbf{r}^\mathbf{T}\mathbf{W}\mathbf{J}$$

Our update rule for gradient descent can then be written as

$$\mathbf{a}_{s+1} = \mathbf{a}_s + \gamma\mathbf{r}^\mathbf{T}\mathbf{W}\mathbf{J}$$

5.7   Implementing gradient descent

Example 5.5. Implement gradient descent with the model data described earlier in this chapter.

Solution 5.5. We load the data and do the calculation

```
%Imprementing gradiate descent
clear;clf;figure(1);
load modeldata.mat

errorbar(xi, baryi, barsigi,'.k');
hold on;
xlabel('time [s]', 'fontsize',20);
```

```matlab
ylabel('distance [m]', 'fontsize',20);

N = length(xi);
W = zeros(N,N);
for i = 1:N
    W(i,i) = 1/barsigi(i)^2;
end
%some initial random guess
A(1) = .2
B(1) = -.2
C(1) = .3


gamma = 0.00001; %Arbitrary step size
S = 10000;   %Number of steps

for s = 1:S
 for i = 1:N
  J(i,1) = exp(-C(s)*xi(i));
  J(i,2) = xi(i)*exp(-C(s)*xi(i));
  J(i,3) = (A(s)+B(s)*xi(i) )*(-xi(i))*exp(-C(s)*xi(i));
  ri(i) = (baryi(i)-(A(s)+B(s)*xi(i))*exp(-C(s)*xi(i)));
 end
    delta = 2*gamma*ri*W*J;
    A(s+1)=A(s)+delta(1);
    B(s+1)=B(s)+delta(2);
    C(s+1)=C(s)+delta(3);
end;

ymodel = (A(end)+B(end).*xi).*exp(-C(end).*xi);
plot(xi, ymodel,'r');
title('Gradient descent fits','fontsize',20);

%The end parameters are
>> A(end)
>> 1.9287
>> B(end)
>> -4.8880
>> C(end)
>> 0.9908
```

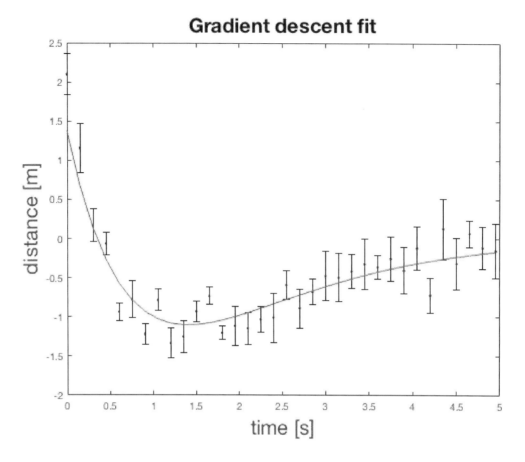

Figure 5.3: Gradient descent works

5.8   Gauss-Newton algorithm

$$r_i = \overline{y}_i - f(x_i; \mathbf{a})$$

The Gauss-Newton algorithm is another iterative curve fitting method. One can arrive at the update formula by considering a Taylor expansion of $f(x_i; \mathbf{a})$ with respect to an update $\boldsymbol{\delta}$.

$$f(x_i; \mathbf{a} + \boldsymbol{\delta}) \approx f(x_i, \mathbf{a}) + \frac{\partial f(x_i; \mathbf{a})}{\partial \mathbf{a}}\boldsymbol{\delta}$$
$$f(x_i; \mathbf{a} + \boldsymbol{\delta}) = \mathbf{f} + \mathbf{J}\boldsymbol{\delta}$$

At the minimum of χ^2, then

$$\frac{\partial \chi^2}{\partial \boldsymbol{\delta}} = 0$$

Then we have

$$\chi^2(\mathbf{a} + \boldsymbol{\delta}) \approx (\mathbf{y} - \mathbf{f}(x_i; \mathbf{a}) - \mathbf{J}\boldsymbol{\delta})^T \mathbf{W}(\mathbf{y} - \mathbf{f}(x_i; \mathbf{a}) - \mathbf{J}\boldsymbol{\delta})$$
$$\chi^2(\mathbf{a} + \boldsymbol{\delta}) \approx (\mathbf{r} - \mathbf{J}\boldsymbol{\delta})^T \mathbf{W}(\mathbf{r} - \mathbf{J}\boldsymbol{\delta})$$

We want to take the derivative of χ^2 with respect to $\boldsymbol{\delta}$ and set it equal to zero. Remember the important derivative.

$$0 = (\mathbf{r} - \mathbf{J}\boldsymbol{\delta})^T \mathbf{W}(-\mathbf{J})$$

We can take the transpose of that expression to get

$$0 = \mathbf{J}^T \mathbf{W}(\mathbf{r} - \mathbf{J}\boldsymbol{\delta})$$

Remember that $\mathbf{W}$ is symmetric. Solving for $\boldsymbol{\delta}$ gives the Gauss Newton update rule.

$$\boldsymbol{\delta} = (\mathbf{J}^T \mathbf{W} \mathbf{J})^{-1} \mathbf{J}^T \mathbf{W} \mathbf{r}$$
$$\mathbf{a}_{s+1} = \mathbf{a}_s + \boldsymbol{\delta}$$

5.9 Implementing Gauss Newton

To implement the Gauss Newton method, we only have to really change one line of code for the update rule. Gauss Newton seemed to be pickier about the initial guess.

Example 5.6. Implement the Gauss Newton method to fit the model data

Solution 5.6. We change the initial guess and update the delta rule.

```
clear;clf;figure(1);
load modeldata.mat

errorbar(xi, baryi, barsigi,'.k');
hold on;
xlabel('time [s]', 'fontsize',20);
ylabel('distance [m]', 'fontsize',20);

N = length(xi);
W = zeros(N,N);
for i = 1:N
    W(i,i) = 1/barsigi(i)^2;
end
%some initial random guess
A(1) = 1.4
B(1) = -4.3
C(1) = 1.3


gamma = 0.00001; %Arbitrary step size
S = 10000;   %Number of steps
```

```
for s = 1:S
    for i = 1:N
        J(i,1) = exp(-C(s)*xi(i));
        J(i,2) = xi(i)*exp(-C(s)*xi(i));
        J(i,3) = (A(s) + B(s)*xi(i) )*(-xi(i))*exp(-C(s)*xi(i));
        ri(i) = (baryi(i) - (A(s)+ B(s)*xi(i))*exp(-C(s)*xi(i)));
    end
    % we really only have to change one line of code for delta
    delta = inv(J'*W*J)*J'*W*ri';
    A(s+1)=A(s)+delta(1);
    B(s+1)=B(s)+delta(2);
    C(s+1)=C(s)+delta(3);
end;

ymodel = (A(end)+B(end).*xi).*exp(-C(end).*xi);
plot(xi, ymodel,'r');
title('Gauss Newton fits','fontsize',20);
```

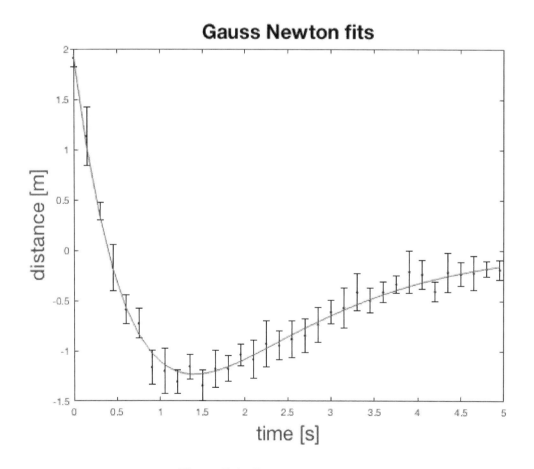

Figure 5.4: Gauss Newton works

5.10 Levenberg Marquadt Method

The Levenberg-Marquardt algorithm is one of the most powerful curve fitting methods available. Most commercial software packages use it in one form or another. It is robust, meaning that it will converge to a local minimum for a wide range of initial guesses. It does not always find the global minimum. The Levenberg-Marquardt algorithm can be thought of as a combination of the Gauss-Newton method and the gradient descent method. To derive the Levenberg-Marquardt algorithm one starts along the same lines as the Gauss-Newton method and we arrive at

$$(\mathbf{J}^T\mathbf{W}\mathbf{J})\delta = \mathbf{J}^T\mathbf{W}\mathbf{r}$$

The Levenberg-Marquardt method slightly modifies this equation to

$$(\mathbf{J}^T\mathbf{W}\mathbf{J} + \lambda\operatorname{diag}(\mathbf{J}^T\mathbf{W}\mathbf{J}))\delta = \mathbf{J}^T\mathbf{W}\mathbf{r}$$

The diag symbol means only the components along the diagonal are nonzero. and the update δ can be solved for by an inverse matrix. The Levenberg-Marquardt update rule is given by

$$\delta = [\mathbf{J}^T\mathbf{W}\mathbf{J} + \lambda\operatorname{diag}\mathbf{J}^T\mathbf{W}\mathbf{J}]^{-1}\mathbf{J}^T\mathbf{W}\mathbf{r}$$
$$\mathbf{a}_{s+1} = \mathbf{a}_s + \delta$$

5.11 Implementing L.M.

Example 5.7. Implement the Levenberg Marquardt algorithm in MATLAB

Solution 5.7. `clear;clf;figure(1);`
```
load modeldata.mat

errorbar(xi, baryi, barsigi,'.k');
hold on;
xlabel('time [s]', 'fontsize',20);
ylabel('distance [m]', 'fontsize',20);

N = length(xi);
W = zeros(N,N);
for i = 1:N
    W(i,i) = 1/barsigi(i)^2;
end
%some initial random guess
A(1) = 1
B(1) = -1
C(1) = 0

S = 1000;
lambda = 0.01;
```

```
for s = 1:S
    for i = 1:N
        J(i,1) = exp(-C(s)*xi(i));
        J(i,2) = xi(i)*exp(-C(s)*xi(i));
        J(i,3) = (A(s) + B(s)*xi(i) )*(-xi(i))*exp(-C(s)*xi(i));
        ri(i) = (baryi(i) - (A(s)+ B(s)*xi(i))*exp(-C(s)*xi(i)));
    end
    Q = J'*W*J;
    Q(1,1) = Q(1,1)*(1+lambda);
    Q(2,2) = Q(2,2)*(1+lambda);
    Q(3,3) = Q(3,3)*(1+lambda);
    delta = inv(Q)*J'*W*ri';
    A(s+1)=A(s)+delta(1);
    B(s+1)=B(s)+delta(2);
    C(s+1)=C(s)+delta(3);
end;

ymodel = (A(end)+B(end).*xi).*exp(-C(end).*xi);
plot(xi, ymodel,'r');
title('Levenberg Marquardt fits','fontsize',20);

>> A(end)
>> 1.9405
>> B(end)
>> -4.9400
>> C(end)
>> 0.9959
```

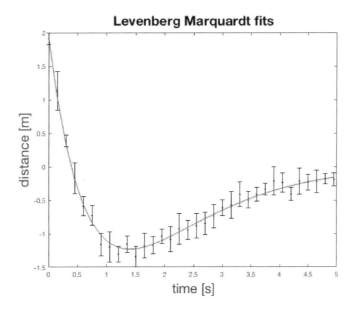

Figure 5.5: Levenberg Marquardt works

5.12 LM parameter uncertainties

We came so far in this chapter and found the best fit parameters. Let's find the best fit parameter uncertainties using the formula we described earlier.

$$D_{jk} = \sum_{i=1}^{N} \frac{1}{\overline{\sigma}_i^2} \left[\left(\frac{\partial f}{\partial a_j} \right) \left(\frac{\partial f}{\partial a_k} \right) + (f - \overline{y}_i) \frac{\partial^2 f}{\partial a_j \partial a_k} \right]$$

Since our model is

$$f = (A + Bt) \exp(-Ct)$$

Example 5.8. Implement the MATLAB code to calculate the Levenberg-Marquardt nonlinear least squares uncertainties.

Solution 5.8. I found the partial derivatives in Mathematica. Let's go to the MATLAB code then.

```
wi = barsigi.^(-2);
A=A(end);
B=B(end);
C=C(end);

E1 = exp(-C.*xi);
E2 = exp(-2.*C.*xi);
P1 = (A + B.*xi);
X2 = xi.*xi;
D(1,1) = sum( wi.*E2);
```

```
D(2,2) = sum( wi.*X2.*E2);
D(3,3) = sum( wi.*(E2.*X2.*P1.^2)-E1.*X2.*P1.*(-E1.*P1 + baryi));
D(1,2) = sum( wi.*E2.*xi);
D(1,3) = sum( -wi.*E2.*xi.*P1 + wi.*E1.*xi.*(-E1.*P1 + baryi));
D(2,3) = sum( -wi.*E2.*X2.*P1 + wi.*E1.*X2.*(-E1.*P1 +baryi));
D(2,1) = D(1,2);
D(3,1) = D(1,3);
D(3,2) = D(2,3);
D
M = inv(D)
A
sqrt(M(1,1))
B
sqrt(M(2,2))
C
sqrt(M(3,3))
```

For a fresh run, I got $A = 2.03(14)$, $B = -5.09(25)$, and $C = 0.989(26)$ which seems reasonable. Be grateful for computers.

5.13   Summary

- Curvature expression for **D**

$$D_{jk} = \sum_{i=1}^{N} \frac{1}{\overline{\sigma}_i^2} \left[\left(\frac{\partial f}{\partial a_j} \right) \left(\frac{\partial f}{\partial a_k} \right) + (f - \overline{y}_i) \frac{\partial^2 f}{\partial a_j \partial a_k} \right]$$

- Nonlinear uncertainties. $\mathbf{C} = \mathbf{D}^{-1}$.

$$\sqrt{C_{jj}} = \overline{\sigma}_{a_j}$$

- Newton's method update rule

$$x_{n+1} = x_n - \frac{f(x_n)}{f'(x_n)}$$

- Gradient descent update rule

$$\mathbf{a}_{s+1} = \mathbf{a}_s + \gamma \mathbf{r}^\mathbf{T} \mathbf{W} \mathbf{J}$$

- Gauss Newton update rule

$$\boldsymbol{\delta} = (\mathbf{J}^T \mathbf{W} \mathbf{J})^{-1} \mathbf{J}^T \mathbf{W} \mathbf{r}$$
$$\mathbf{a}_{s+1} = \mathbf{a}_s + \boldsymbol{\delta}$$

- Levenberg Marquardt update rule

$$\boldsymbol{\delta} = [\mathbf{J^T W J} + \lambda \text{diag} \mathbf{J^T W J}]^{-1} \mathbf{J^T W r}$$
$$\mathbf{a}_{s+1} = \mathbf{a}_s + \boldsymbol{\delta}$$

Chapter 6

Introduction to Probability and Statistics

6.1 Basic concepts of probability

6.1.1 Sum rule and product rule for independent events

One can understand some basic facts about probability by considering rolls of dice. If you have a die, then the probability of rolling any one side is one sixth, $P(k) = 1/6$. The probability to roll one side versus another is independent if the die is fair. In error analysis, we roll the die over and over again and collect data. We try to see if the die is fair. In probability, we just assume it is perfect and derive the consequences.

If you want to calculate the probability of rolling an even number, then you add the probabilities.

$$P_{even} = P(2) + P(4) + P(6) = 1/6 + 1/6 + 1/6 = 1/2$$

The other important rule is when you have independent successive events. For example, the probability of rolling a one twice in a row is

$$P_{1,1} = (1/6)(1/6) = 1/36$$

The fact that you may have rolled a one on the first roll does not influence what happens on the second roll.

We can sumarize these results for independent events as

$$P(A \, or \, B) = P(A) + P(B)$$

$$P(A \, and \, then \, B) = P(A)P(B)$$

6.1.2 Combinatorics

For our discussion or probability, we need some basic notions of combinatorics. If there are N different objects, then the number of ways you can permute or rearrange the objects is given by

$$_N P_N = N!$$

Example 6.1. You have an apple, a banana, an orange, and a watermelon. How many different orders can eat the fruit in?

Solution 6.1. This is a standard permutation problem. There are four different objects. If you choose one of the four to eat first there are four ways to do that. Then there are 3 ways to choose the next and then 2 ways to choose the next and one way to choose the last. The total number of permutations is

$$P_4 = 4 \times 3 \times 2 \times 1 = 4!$$

Another common combinatorics problem is the number of ways you can choose n objects from N. The number of combinations is given by the binomial coefficient.

$$_N C_n = \binom{N}{n} = \frac{N!}{n!(N-n)!}$$

Example 6.2. There are ten players on the basketball team. How many different ways can you choose the starting line-up of 5 players?.

Solution 6.2. You are asked to calculate $\binom{10}{5}$ so we have.

$$\binom{10}{5} = \frac{10!}{5!5!} = \frac{10 \times 9 \times 8 \times 7 \times 6}{5 \times 4 \times 3 \times 2 \times 1} = 9 \times 7 \times 4 = 252$$

6.1.3   Sterling formula

Factorial is a special calculation. Computing the factorial of large numbers on your calculator is generally difficult because there will be more digits required than available than on your screen. There is a convenient approximate formula for the factorial of large numbers called Stirling's formula.

$$n! = \sqrt{2\pi n}\, n^n e^{-n}$$

Factorial in general, can be represented as an integral.

$$n! = \int_0^\infty t^n e^{-t} dt$$

This allows you to find even factorials for fractions or certain negative numbers. When one says $n!$ one is generally talking about nonnegative integers. If you want to have other numbers, we use the gamma function.

$$\Gamma(z+1) = \int_0^\infty t^z e^{-z} dz \quad \Gamma(z+1) = z\Gamma(z)$$

One also often works with the natural logarithm of the factorial function. Then

$$\ln(n!) = \ln(\sqrt{2\pi}) + \frac{1}{2}\ln n + n\ln n - n$$

6.2 Probability distributions

6.2.1 Discrete probability distirbutions

If there a finite number of possible outcomes we can describe the expected frequency of events by a discrete probability distribution. If there are n different possibilities, then we can define the probabilities

$$p_k \quad k = 1, 2, \ldots n$$

p_k is just the number of times you expect k to occur out of N total measurements. The probabilities have to be nonnegative.

$$p_k \geq 0$$

The probabilities have to add up to 1. This is called the normalization condition.

$$\sum_{k=1}^{n} p_k = 1$$

Example 6.3. What is the probability distribution for rolling a fair die?

Solution 6.3. Let p_k be the probability of rolling k on a fair die, then

$$p_k = 1/6 \quad k = 1, 2, 3, 4, 5, 6$$

Each side is equally likely and has the same probability. The sum of the probabilities is 1. This distribution is also called a uniform distribution.

Some important discrete distributions we will consider in this chapter are the binomial distribution and the Poisson distribution.

6.2.2 Continuous probability distributions

Other quantities are best described by a continuous distribution. For example, height or intelligence are continuous quantities. These are recorded as discrete numbers however, but that is an artefact of the measurement recording. The probability for a measurement to lie between y and $y + dy$ is

$$\rho(y)dy$$

The probability for a measurement Y to lie between a and b is

$$P(a \leq Y \leq b) = \int_a^b \rho(y)dy$$

The probability distribution must be nonnegative.

$$\rho(y) \geq 0 \quad all\, y$$

The probability must add up to one.

$$\int_{-\infty}^{\infty} \rho(y)dy = 1$$

The most important continuous probability distribution is the gaussian distribution which is also called the normal distribution.

6.2.3   Frequency distributions and the limiting distribution

We can take a set of measurements such as the age in years of students in the class. We can place them into bins say one year wide. Then we could make a bar graph of the different ages. This is called a histogram. The frequency distribution is the number in each bar divided by the total number of measurements. The sum of the frequencies will add up to one. The frequencies are the experimental version of the probability distribution.

If you take a set of N measurements you might get some idea of the distribution of events, but there will be inherent randomness. In the limit of an infinite number of measurements, then you reach what is called the limiting distribution. Often the idea is that the limiting distribution will be some known function because you understand the nature of how the measurements arise.

6.3   Expectation value

6.3.1   Discrete expectation value

If you have a discrete probability distribution the expectation value of a function $f(y)$ is defined as

$$E[f(y)] = \sum_{k=1}^{n} f(y_k)p_k$$

You would use the frequency distribution you measure to calculate the quantity. In the theoretical limit of infinite data, you can understand what a limiting distribution will produce for comparison.

6.3.2   Continuous expectation value

f you have a continuous probability distribution $\rho(y)$ the expectation value of a function $f(y)$ is defined as

$$E[f(y)] = \int_{-\infty}^{\infty} f(y)\rho(y)dy$$

If you are worried about the infinity, it occurs that the probability is zero or negligible near there.

6.3.3   Moments

The kth moment is defined as

$$E[y^k] = \sum_{k=1}^{n} y^k p_k$$

We are usually interested in the mean which is

$$E[y] = \mu = \bar{y} = \sum_{k=1}^{n} k p_k$$

The other important statistic to calculate is standard deviation. Recall the formula for variance.

$$Variance = E[(y - E[y])^2] = E[y^2] - E[y]^2$$

Since the standard deviation is the square root of the variance we just need to compute the first two moments to find it in probability theory.

6.3.4 Using moment generating functions

We define the moment generating function as

$$M_Y(t) = E[e^{ty}] \quad t \in R$$

If we take the expectation value of $M_Y(t)$ by Taylor series we find.

$$E[e^{ty}] = E[1] + E[yt] + \frac{E[(yt)^2]}{2!} + \dots$$

From this formula we see the kth moment is just

$$E[y^k] = \frac{d^k}{dt^k} E[e^{ty}](0)$$

Example 6.4. The binomial distribution is defined as

$$b(y; n, p) = \frac{n!}{x!(n-y)!} p^y q^{n-y} \quad q = 1 - p$$

The binomial distribution is the probability of measuring 1, y times with a probability p each, and 0, $n - y$ times with a probability $(1 - p)$ each. It derives from the binomial expansion

$$(a + b)^n = \sum_{k=0}^{n} \binom{n}{k} a^k b^{n-k}$$

Find the moment generating function, the mean, and the standard deviation

Solution 6.4. The moment generating function is

$$M(t) = \sum_{y=0}^{n} e^{yt} \frac{n!}{x!(n-y)!} p^y q^{n-y} = \sum_{y=0}^{n} \frac{n!}{x!(n-y)!} (pe^t)^y q^{n-y} = (q + pe^t)^n$$

The first moment is given by

$$M'(t) = n(q + pe^t)^{n-1} pe^t \quad E[y] = M'(0) = np$$

The second moment is given by

$$
\begin{aligned}
M''(t) &= npe^t[(n-1)(q+pe^t)^{n-2}pe^t] + (q+pe^t)^{n-1}(npe^t) \\
&= npe^t(q+pe^t)^{n-2}[(n-1)pe^t + (q+pe^t)] \\
&= npe^t(q+pe^t)^{n-2}[q+npe^t]
\end{aligned}
$$

$$
M''(0) = np(q+np) \quad E[y^2] - E[y]^2 = np(q+np) - (np)^2 = npq
$$

For the binomial distribution, the mean is np and the standard deviation is $\sqrt{npq}$.

Example 6.5. The normal distribution is defined as

$$
N(y;\mu,\sigma) = \frac{1}{\sigma\sqrt{2\pi}}\exp[-(y-\mu)^2/(2\sigma^2)]
$$

Find the moment generating function, the mean, and the standard deviation

Solution 6.5. The moment generating function is given as

$$
M(t) = \int_{-\infty}^{\infty} \frac{e^{ty}}{\sigma\sqrt{2\pi}}\exp[-(y-\mu)^2/(2\sigma^2)]dx
$$

To do this integral let $z = (y-\mu)/\sigma$. Then we calculate.

$$
M(t) = e^{\mu t}\int_{-\infty}^{\infty} \frac{e^{z\sigma t}e^{-\frac{1}{2}z^2}}{\sqrt{2\pi}}dz = e^{\mu t}\int_{-\infty}^{\infty} \frac{e^{-\frac{1}{2}(z^2+2z\sigma)}}{\sqrt{2\pi}}
$$

If you complete the square you find.

$$
M(t) = e^{\mu t+\frac{1}{2}\sigma^2 t^2}\int_{-\infty}^{\infty} \frac{e^{-\frac{1}{2}z^2}}{\sqrt{2\pi}} = e^{\mu t+\frac{1}{2}\sigma^2 t^2}
$$

That integral is just equal to one. We will discuss some basic gaussian integrals to derive this result later.

$$
M(t) \approx (1+\mu t+(1/2)\mu^2 t^2+\ldots)(1+1/2\sigma^2 t^2+\ldots)
$$

$$
M(t) \approx (1+\mu t+(1/2)\sigma^2 t^2+(1/2)\mu^2 t^2+\ldots)
$$

Since $M'(0) = \mu$ the mean is μ and since $M''(0) = \sigma^2 + \mu^2$. The variance is given by

$$
\sigma^2 = M''(0) - M'(0)^2 = \sigma^2
$$

The standard deviation is just σ.

Example 6.6. The Poisson distribution is defined as

$$p_k = \frac{\lambda^k e^{-\lambda}}{k!}$$

Find the moment generating function, the mean, and the standard deviation

Solution 6.6. $M(t)$ is given by the sum

$$M(t) = \sum_{k=0}^{\infty} e^{tk} \frac{\lambda^k e^{-\lambda}}{k!} = e^{-\lambda} \sum_{k=0}^{\infty} \frac{(\lambda e^t)^k}{k!} = e^{\lambda(e^t - 1)}$$

Now we can easily find the mean and the second moment

$$M'(t) = M(t)\lambda e^t \quad E[y] = M'(0) = \lambda$$
$$M''(t) = M'(t)\lambda e^t + M(t)\lambda e^t = M(t)[\lambda^2 e^{2t} + \lambda e^t]$$

$$E[y^2] = M''(0) = \lambda^2 + \lambda \quad \sigma^2 = E[y^2] - E[y]^2 = \lambda^2 + \lambda - \lambda^2 = \lambda$$

So our results are that the mean is λ and the standard deviation is $\sqrt{\lambda}$. The uncertainty being the square root of the mean is often referred to as the square root counting rule for Poisson processes like radioactivity.

6.4 Normal distribution

6.4.1 Gaussian integrals

Gaussian functions are variations of e^{-y^2}. Often in probability and statistics you take integrals of these functions which are called gaussian integrals. When working with gaussian integrals it is important to know some mathematical results.

Example 6.7. Evaluate

$$I = \int_{-\infty}^{\infty} e^{-y^2} dy$$

Solution 6.7. There is no standard integration rule to calculate this. There is a trick you can use however.

$$I^2 = \int_{-\infty}^{\infty} \int_{-\infty}^{\infty} e^{-x^2 - y^2} dx dy$$

Now you can convert to polar coordinates where $dx dy = r dr d\theta$ and $x^2 + y^2 = r^2$.

$$I^2 = \int_0^{2\pi} \int_0^{\infty} e^{-r^2} r dr d\theta = (2\pi)(1/2) = \pi$$

So we have

$$I = \int_{-\infty}^{\infty} e^{-x^2} dx = \sqrt{\pi}$$

It is worth noting that we can do other similar integrals by a change of variable

$$\int_{-\infty}^{\infty} e^{-\alpha y^2} dy = \sqrt{\frac{\pi}{\alpha}}$$

Example 6.8. Evaluate

$$\int_{-\infty}^{\infty} x e^{-x^2} dx \qquad \int_{-\infty}^{\infty} x^2 e^{-x^2}$$

Solution 6.8. The first integrand is odd so

$$\int_{-\infty}^{\infty} x e^{-x^2} dx = 0$$

The left of zero area exactly cancels the right of zero area. The second integral we can find by taking a partial derivative.

$$\int_{-\infty}^{\infty} e^{-\alpha x^2} dx = \sqrt{\frac{\pi}{\alpha}}$$

Take the partial derivative of both sides with respect to $-\dfrac{\partial}{\partial \alpha}$. Then we get the formula

$$\int_{-\infty}^{\infty} x^2 e^{-\alpha x^2} dx = \frac{1}{2\alpha}\sqrt{\frac{\pi}{\alpha}}$$

Then set $\alpha = 1$ and we have

$$\int_{-\infty}^{\infty} x^2 e^{-x^2} = \frac{\sqrt{\pi}}{2}$$

6.4.2   Definition of the Gaussian distribution

$$G(x; \mu, \sigma) = \frac{1}{\sqrt{2\pi\sigma^2}} e^{-(x-\mu)^2/2\sigma^2}$$

We can see that the Gaussian distribution is normalized with our results from gaussian integrals. From our results with the moment generating functions, we have that μ is the mean and σ is the standard deviation of the normal distribution or Gaussian distribution. Normal distributions occur when you have a large sum of independent factors determining the distribution.

6.4.3   Confidence intervals

If your data follow a gaussian distribution how confident can you be that a single measurement will fall close to the mean. The probability to fall within $y \in (\mu - \sigma, \mu + \sigma)$ is 68 percent. This is calculated from the area under the curve about the mean.

Example 6.9. Find the confidence intervals about the mean for σ, 2σ, 3σ, 4σ, and 5σ.

Solution 6.9. The confidence interval for different uncertainties is defined as the probability for a measurement to lie between $\pm n\sigma$. We can calculate this as a definite integral numerically. For example, the one sigma confidence interval is

$$0.68 = \int_{\mu-\sigma}^{\mu+\sigma} \frac{1}{\sigma\sqrt{2\pi}} e^{-(x-\mu)^2/(2\sigma^2)} dx$$

$$0.95 = \int_{\mu-2\sigma}^{\mu+2\sigma} \frac{1}{\sigma\sqrt{2\pi}} e^{-(x-\mu)^2/(2\sigma^2)} dx$$

and so on when you calculate the integrals numerically. About the mean, for 3σ we have 0.997, for 4σ we have 0.99994, and for 5σ we have 0.9999994. One sees that if the data follows a normal distribution it is highly unlikely a measurement will wander more than 5σ away from the mean. You wouldn't predict that.

6.5 Poisson distribution

6.5.1 Meaning of the Possion distribution

When things occur at a constant rate but randomly their distribution is described by a Poisson distribution. Some example include the number of emails you get on Mondays, the number of radioactive decays in a 10 second interval, or the number of chocolate chips in your cookies.

6.5.2 Definition of the Poisson distribution

The Poisson distribution is defined as

$$p_k = \frac{\lambda^k e^{-\lambda}}{k!}$$

6.5.3 Normalization of the Poisson distribution

Recall that

$$e^x = 1 + \frac{x}{1!} + \frac{x^2}{2!} + \frac{x^3}{3!} + \dots$$

The sum of p_k is easy to calculate.

$$\sum_{k=0}^{\infty} p_k = \sum_{k=0}^{\infty} \frac{\lambda^k e^{-\lambda}}{k!} = e^{-\lambda} \sum_{k=0}^{\infty} \frac{\lambda^k}{k!} = e^{-\lambda} e^{\lambda} = 1$$

6.5.4   Mean of the Poisson distribution

The mean of the Poisson distribution is

$$E(k) = \sum_{k=0}^{\infty} k \frac{\lambda^k e^{-\lambda}}{k!} = \sum_{k=1}^{\infty} \frac{\lambda^k e^{-\lambda}}{(k-1)!} = \lambda \sum_{k=1}^{\infty} \frac{\lambda^{k-1} e^{-\lambda}}{(k-1)!} = \lambda \sum_{k=0}^{\infty} \frac{\lambda^k e^{-\lambda}}{k!} = \lambda$$

We also found the same result using the moment generating function.

6.5.5   The standard deviation of the Poisson distribution

The second moment of the Poisson distribution is

$$E(k^2) = \sum_{k=1}^{\infty} k \frac{\lambda^k e^{-\lambda}}{(k-1)!} = \sum_{k=1}^{\infty} (k-1+1) \frac{\lambda^k e^{-\lambda}}{(k-1)!} = \lambda + \sum_{k=2}^{\infty} \frac{\lambda^k e^{-\lambda}}{(k-2)!} = \lambda + \lambda^2$$

The standard deviation of the Poisson distribution is therefore

$$\sigma^2 = E(k^2) - E(k)^2 = \lambda \qquad \sigma = \sqrt{\lambda}$$

This is the proof of the square root counting rule.

> **Example 6.10.** You have enough cookie dough for 100 equal cookies. You add 500 chocolate chips. What is the probability that a cookie has 0 chocolate chips, 5 chocolate chips or 7 chocolate chips?
>
> **Solution 6.10.** The average number of chocolate chips per cookie is $\lambda = 5$.
>
> $$p_0 = \frac{5^0 e^{-5}}{0!} = 0.0067$$
>
> $$p_5 = \frac{5^5 e^{-5}}{5!} = 0.175$$
>
> $$p_7 = \frac{5^7 e^{-5}}{7!} = 0.104$$

6.5.6   The gaussian approximation to the Poisson distribution

$$P_\lambda(k) = \frac{1}{\sqrt{2\pi\lambda}} \exp\left[-\frac{(k-\lambda)^2}{2\lambda} \right]$$

6.6   Binomial distribution

6.6.1   Coin tossing

If you toss a coin say a 100 times you would expect the number of heads to be close to the number of tails, 50 each. The probability of n heads out of N tosses is given by the binomial distribution. q is defined as $1 - p$.

$$P(n|N) = \binom{N}{n}(1/2)^n(1 - 1/2)^{N-n} = \binom{N}{n}/2^N$$

If the coin is not fair then the binomial distribution for n successes out of N is given by

$$P(n|N,p) = \binom{N}{n}p^n(1 - p)^{N-n}$$

There is also a normal approximation to the binomial distribution when N is large given by

$$\rho(n|N,p) \approx \frac{1}{\sqrt{2\pi Npq}}\exp\left[-\frac{(n - Np)^2}{2Npq}\right]$$

6.6.2 Unbalanced games

Example 6.11. Suppose you play a gambling game with a house advantage of 3 percent. If you have a 1000 dollars to play, is it better to just make one big bet or to split the bet up into 1000 one dollar bets? Your goal is to win money. What is the probability to win money, break even, and lose. How much do you expect to win or lose?

Solution 6.11. We can calculate the probabilities to win, tie, and lose.

$$P_{win} = \sum_{k=501}^{1000}\binom{1000}{k}(0.47)^k(0.53)^{1000-k} = 0.9691$$

$$P_{lose} = \sum_{k=0}^{499}\binom{1000}{k}(0.47)^k(0.53)^{1000-k} = 0.0267$$

$$P_{tie} = \binom{1000}{500}(0.47)^{500}(0.53)^{500} = 0.0042$$

If you make $N = 1000$ bets with a probability $p = 0.47$ of winning you expect to win

$$M = M_+p + M_-(1 - p) = (2000)(0.47) + (0)(0.53) = 940$$

So you expect to lose 60 dollars. One can also calculate this from the binomial distribution.

$$M = \sum_{k=0}^{N}2k\binom{N}{k}p^k(1 - p)^{N-k} - 1000 = 940$$

```
E = 0; N = 1000;
for k = 1:N
E = E + 2*k*bincoeff(N,k)*(0.47)^k*(0.53)^(N-k);
end
E = E - N;
```

We can break this sum up into two parts, above 500 wins versus tying or losing. $E_+ = 27.08$ and $E_- = 912.92$. This basically says when you don't win for sure, you lose \$97.08. But when you win for sure you only get about \$27.08. Not balanced at all, and only a 2.67 percent chance to win.

$$912.92 + 27.08 = 940$$

You only have a realistic chance of winning if you make one large bet with a probability of 47 percent. The downside is that if you lose then all your money is gone. Casinos want you to split up your bets because on average it is more likely you lose the longer you play. Making a 1000 bets is probably also not worth your time to win only 27 bucks.

As an exercise, try to set up the probabilities as integrals under the normal approximation for comparison. This is beyond the basic features of MATLAB without numerical analysis. It is best to do the integrals in Mathematica.

6.7   Maximum likelihood

Suppose that you hypothesize your data follows a certain theoretically known limiting distribution. The method of maximum likelihood provides a way to calculate the parameters in these statistical formulas from your dataset. Let y_i be your measurements and a_i be the parameters in the distribution. We will consider the normal distribution, Poisson distribution, and exponential distribution as examples. The normal distribution has two parameters μ and σ. The Poisson distribution and the exponential distribution have one parameter λ.

The maximum likelihood formula is defined as

$$L(y_1, y_2, \ldots y_n | \mathbf{a}) = P(y_1)P(y_2) \ldots P(y_n)$$

The optimization formulas

$$\frac{\partial L}{\partial a_j} = 0$$

give formulas for the parameters a_j. Often it is easier to find the optimized formulas by using $\ln L$ instead of L which doesn't change the results.

6.7.1   Normal distribution

Example 6.12. Calculate the best estimate for μ and σ for the normal distribution using maximum likelihood.

Solution 6.12. The maximum likelihood function for the normal distribution is given by this formula.

$$L(y_1, y_2, \ldots y_n | \mu, \sigma) = \frac{1}{\sigma^n (\sqrt{2\pi})^n} \exp\left[\sum_{i=1}^{n} \frac{(y_i - \mu)^2}{2\sigma^2} \right]$$

Now we take the natural logarithm.

$$\ln L = -(n/2)\ln(2\pi) - n\ln\sigma - \frac{\sum_{i=1}^{n}(y_i - \mu)^2}{2\sigma^2}$$

To find the best estimate of μ we calculate $\dfrac{\partial \ln L}{\partial \mu} = 0$ This formula gives

$$\frac{\sum_{i=1}^{n}(y_i - \mu)}{\sigma^2} = 0$$

This is the familiar formula for the arithmetic mean.

$$\mu = \frac{1}{n}\sum_{i=1}^{n} y_i$$

If one calculates $\dfrac{\partial \ln L}{\partial \sigma} = 0$ one finds

$$-n/\sigma + \frac{\sum_{i=1}^{n}(y_i - \mu)^2}{2\sigma^3} = 0$$

One then finds

$$\sigma = \sqrt{\frac{1}{n}\sum_{i=1}^{n}(y_i - \mu)^2}$$

This is the familiar formula for the standard deviation.

6.7.2 Poisson distribution

Example 6.13. Calculate the best estimate for λ for the Poisson distribution using maximum likelihood.

Solution 6.13. We can do maximum likelihood for the Poisson distribution.

$$L(y_1, y_2, \ldots, y_n|\lambda) = \frac{\lambda^{y_1} e^{-\lambda}}{y_1!} \cdots \frac{\lambda^{y_n} e^{-\lambda}}{y_n!}$$

$$\ln(L) = -n\lambda + \ln(\lambda)\sum_{k=1}^{n} y_k - \ln(\Pi_{k=1}^{n} y_k!)$$

$$\frac{d(lnL)}{d\lambda} = -n + (1/\lambda)\sum_{k=1}^{n} y_k = 0$$

$$\lambda = \frac{\sum_{k=1}^{n} y_k}{n}$$

We know that λ is the mean for the Poisson distribution according to expectation value also so this makes sense.

6.7.3   Exponential distribution

The exponential distribution is defined as

$$\rho(y; \lambda) = \begin{cases} \lambda e^{-\lambda y} & y \geq 0 \\ 0 & y < 0 \end{cases}$$

Example 6.14. Apply maximum likelihood for N measurements drawn from the same exponential distribution to calculate λ.

Solution 6.14. We have

$$L(y_1, y_2, \ldots, y_n | \lambda) = \lambda^n \exp\left[-\sum_{k=1}^{n} \lambda y_k \right]$$

We take the natural logarithm to simplify the calculation.

$$\ln L = n \ln \lambda - \lambda \sum_{k=1}^{n} y_k$$

Optimizing for lambda we have

$$\frac{d \ln L}{d\lambda} = n/\lambda - \sum_{k=1}^{N} y_k = 0$$

Rearranging

$$\frac{1}{\lambda} = \frac{1}{n} \sum_{k=1}^{n} y_k$$

If you take the expectation value of the first moment of an exponential distribution you get

$$E[y] = \int_0^{\infty} \lambda y e^{-\lambda y} dy = \frac{1}{\lambda}$$

In light of this, our result makes sense for the maximum likelihood.

6.8   Chi squared test

We have had a little flavor for probability distributions that commonly occur like the gaussian, Poisson, and binomial distributions. To test if your data match or support a particular probabilistic theory we compute a chi squared statistic. This is similar to what we encountered before, but the formula is slightly different.

> **Example 6.15.** Generate 100 simulated measurements of radioactive decays within a one-minute interval. The average rate should be 3.2 counts per minute.
>
> **Solution 6.15.** We can generate Poisson random numbers in MATLAB using the randp command.
>
> ```
> Y = [];
> for i = 1:100
> Y(i) = randp(3.2);
> end
> Y
> >>Y =
> ```
>
> ```
>     3   4   3   7   6   2   3   6   2   3   5   5   2   1
>     2   2   5   4   1   3   4   5   3   2   1   2   3
>     2   2   3   1   3   4   3   5   7   5   1   1   5   3
>     3   2   2   4   4   3   3   2   4   4   2   1   4
>     3   5   4   7   4   4   4   1   4   3   3   4   3   6
>     5   3   6   6   1   6   3   2   4   2   5   7   3
>     3   3   3   3   2   6   4   3   1   1   3   5   5   4
>     2   2   1   1   3
> ```
>
> ```
> % Now we can find the number of each
> for j = 1:7
> N(j) = length(find(Y==j))
> end
> %This accounts for all the data for this run
> N =     13 18 28 18 12 7 4
> sum(N) == 100 %is true
> ```
>
> The data consists of ones, twos, ..., and sevens.

> **Example 6.16.** For the simulated data, calculate the mean and the expected number of each possible value. Group the data into bins.
>
> **Solution 6.16.** We want to account for all the possibilities, but some have zero frequency. One if often better off if you group bins with small numbers into a single bin. We bin 0s and 1s together because there are only a few there. We bin 7s or higher together also because there are no events with 8 or more. We can calculate the probabilities for each of these bins and how many we expect in each.
>
> ```
> %The mean is supposed to be 3.2 but the data is random
> L = mean(Y) % Actually L = 3.35 for this dataset
> ```

```
p1 = exp(-L) + L*exp(-L);
p2 = L^2*exp(-L)/factorial(2);
p3 = L^3*exp(-L)/factorial(3);
p4 = L^4*exp(-L)/factorial(4);
p5 = L^5*exp(-L)/factorial(5);
p6 = L^6*exp(-L)/factorial(6);
p7 = 1-p1-p2-p3-p4-p5-p6;
p = [p1 p2 p3 p4 p5 p6 p7]'
```

Bin number	Observed	Theoretically expected
0 or 1	13	15.3
2	18	19.7
3	28	22.0
4	18	18.4
5	12	12.3
6	7	6.9
7 or more	4	5.4

To continue on with the analysis we need to define the chi squared statistic.

$$\chi^2 = \sum_{k=1}^{n} \frac{(O_k - E_k)^2}{E_k}$$

In general, if the difference between the observed values and the theoretical values are small it supports the fact that the test distribution is compatible. Let's calculate for our 7 bins.

```
E = 100*p;
O = N;
chi2 = (O - E).^2./E
sum(chi2)
```

$$\chi^2 = \sum_{k=1}^{7} \frac{(O_k - E_k)^2}{E_k} = 2.5$$

The statisticians now say the following. When $\chi^2 \gg N$ where N is the number of bins there is poor agreement of the data with the theoretical distribution. Since we have $n = 7$ bins clearly this is consistent for the data being a Poisson distribution. When you have less measurements there is more chance that statistical anomalies occur and you can't tell if the distribution matches. Also if you have large measurement errors it might cloud this test. It is easier to rule out than it is to conclude. This test either rules it out with a certain criteria or says the result is consistent with the original hypothesis. It is important to point out it doesn't prove the hypothesis, just that it is consistent.

To go deeper into the statistics you need to consider the degrees of freedom, and the probability that you would generate a given value for chi squared with those degrees of freedom. You can look up these numbers in tables or calculators on the internet. The degrees of freedom is the number of bins minus the constraints. We have 1 constraint that the number of data points adds up to 100. We have another constraint that the data have a mean for the Poisson distribution. We used this to calculate the probabilities. If we were testing a normal distribution, we would need to calculate the mean and the standard deviation. Then we would have to subtract two degrees of freedom.

If you know the number of constraints, then you can also work with the reduced chi squared statistic.

$$\widetilde{\chi}_d^2 = \frac{\chi^2}{d} \qquad d = N - c$$

N is the number of bins and c is the number of constraints. The number of degrees of freedom has to be at least one more than the constraints. So we wouldn't have grouped our earlier data into 2 bins instead of 7. For our simulation,

$$\widetilde{\chi}_d^2 = 2.5/5 = 0.5$$

As the reduced chi squared gets larger than 1 it is more likely you have a different distribution or have to experimentally rule out the hypothesis of a Poisson distribution. One usually sets the criteria ahead of time that the probability of the reduced chi squared being greater than a certain value is five percent. This also is looked up in tables. For a more in depth discussion, one should consult a statistics book.

6.9 Conclusion

We have now reached the end of the book. I hope you have learned some good methods to analyze your data and found it enjoyable. Hopefully it didn't take you too long to read through and understand the formulas. In the future, I may add some more sections depending on how popular this book is. Please leave a review on Amazon because I would appreciate your feedback. You can easily reach the page to review the book by following this link. http://amzn.to/2wH9eY0

I have tried to cover most of the basics and what you should focus on first. The next challenge is to learn more about probability, statistics, and solidify your MATLAB skills. I have listed some good references you might consider studying.

Chapter 7

References

7.1 Recommended books on error analysis

1. John Taylor. Introduction to error analysis, the study of uncertainties in physical measurements. 1997.

2. Ifan Hughes and Thomas Hase. Measurements and their Uncertainties: A practical guide to modern error analysis. 2010

3. Philip Bevington and D. Keith Robinson. Data Reduction and Error Analysis for the Physical Sciences. 2002

7.2 Recommended books on probability

1. David J. Morin. Probability: For the Enthusiastic Beginner. 2016

2. Steven J. Miller. The Probability Lifesaver: All the Tools You Need to Understand Chance. 2017

3. Dimitri P. Bertsekas and John N. Tsitsiklis. Introduction to Probability, 2nd Edition. 2008

4. E. T. Jaynes and G. Larry Bretthorst . Probability Theory: The Logic of Science. 2003

7.3 Recommended books on statistics

1. Christopher D Barr and Cetinkaya-Rundel, Mine. OpenIntro Statistics: Third Edition. 2015

2. William H. Press and Saul A. Teukolsky. Numerical Recipes 3rd Edition: The Art of Scientific Computing. 2007

7.4   Recommended books on MATLAB

1. Stormy Attaway. Matlab, Fourth Edition: A Practical Introduction to Programming and Problem Solving. 2016

7.5   Recommended books on calculus

1. Jack Merrin. Calculus Power-up. 2009

2. Adrian Banner. The Calculus Lifesaver. 2007

3. Joel R. Hass and Christopher E. Heil. Thomas' Calculus (14th Edition). 2017

Made in the USA
San Bernardino, CA
09 December 2018